Glaciers of the Himalayas

SOUTH ASIA DEVELOPMENT FORUM

Glaciers of the Himalayas

Climate Change, Black Carbon, and Regional Resilience

Muthukumara Mani
Editor

WORLD BANK GROUP

1 2 3 4 24 23 22 21

ISBN (paper): 978-1-4648-0099-3
ISBN (electronic): 978-1-4648-1256-9
DOI: 10.1596/978-1-4648-0099-3

Cover image: ©Anna Kurzaeve / Getty Images. Used with permission of Anna Kurzaeve / Getty Images. Further permission required for reuse.
Cover design: Sergio Andres Moreno Tellez, GCS Creative Services, World Bank.

Library of Congress Control Number: LOC 9781464800993

South Asia Development Forum

Home to a fifth of mankind, and to almost half of the people living in poverty, South Asia is also a region of marked contrasts: from conflict-affected areas to vibrant democracies, from demographic bulges to aging societies, from energy crises to global companies. This series explores the challenges faced by a region whose fate is critical to the success of global development in the early 21st century, and that can also make a difference for global peace. The volumes in it organize in an accessible way findings from recent research and lessons of experience, across a range of development topics. The series is intended to present new ideas and to stimulate debate among practitioners, researchers, and all those interested in public policies. In doing so, it exposes the options faced by decision makers in the region and highlights the enormous potential of this fast-changing part of the world.

Contents

Boxes

Figures

Maps

Tables

Foreword

Climate change is altering the natural water cycles in South Asia. The consequences of melting glaciers diminish seasonal snow, and changes in precipitation pose significant risks to the stability of water resources in the region. Glaciers have been crucial to the balance of the ecosystem: they help to moderate flows in the region's major rivers by providing a source of meltwater in hot, dry years and storing water during colder, wetter years. The Himalaya, Karakoram, and Hindu Kush mountain ranges of South Asia, which contain almost 55,000 glaciers, store more freshwater than anywhere but the North and South Poles.

The recent series of flash floods that killed dozens of people and left hundreds missing in the Himalayas was far from the first such disaster to occur among the world's high-mountain glaciers. In a world with a changing climate, it will not be the last. With the shrinking and thinning of glaciers, it is clear that communities and entire sectors of industry are being—and will continue to be—impacted by the changes in water supply. Farmers and indigenous peoples will be hit the hardest as droughts, floods, and weather fluctuations disrupt crop production and alter the livelihoods of those who live downstream. Many of the 750 million people living in the basins of the three mighty rivers of the region—the Indus, the Ganges, and the Brahmaputra—derive their livelihoods primarily from agriculture, which, in turn, feeds the region's urban communities. Water from the headwaters of these rivers contributes significantly to energy production and recreation in the region.

Recent evidence suggests that in addition to changing temperatures and precipitation patterns, anthropogenic black carbon (BC) deposits—soot—are further accelerating glacier and snow melt in these mountain ranges. BC is generated by human activity both inside and outside South Asia. It is part of a larger basket of aerosols that impact climate change directly and indirectly. A key finding of *Glaciers of the Himalayas* is that fully implementing current BC emissions policies in South Asia can reduce BC deposition in the region by almost one quarter. BC emissions can be reduced by an additional 50 percent by enacting and putting in place new policies that are currently economically and technically feasible.

Glaciers of the Himalayas provides new evidence on the extent to which BC reduction policies by South Asian countries have an impact on glacier formation and melt in the Himalaya, Karakoram, and Hindu Kush mountain ranges within the context of the changing global climate. It also examines the extent of water resources and the potential impact of this loss of glaciation on downstream river basins. The study presents scenarios through to 2040 to align with a reasonable policy-making time horizon.

The report concludes that managing BC emissions in South Asia has the potential not only to achieve global and regional climate benefits, but also to offer other valuable advantages for the region. For example, cleaner cooking and fuel burning, in addition to reducing BC deposits on glaciers, would improve local air quality; help achieve long-term energy security, especially after switching to solar and other clean energy solutions; and ultimately help meet climate targets.

The World Bank Group is stepping up its efforts to help South Asia address the challenges raised by the region's vanishing glaciers. The World Bank has recently established a US$500 million Clean Cooking Fund to galvanize political commitment and investment to achieve universal access to modern energy cooking services (MECS) by 2030. This would indirectly serve the purpose of investing in solutions to reduce BC. Through our recently approved Climate Change Action Plan, we are providing financing and helping develop strategic policies to step up climate action.

Regional cooperation is critical in helping countries in the region collaboratively manage glaciers and other natural assets. This is possible only if countries can exchange information about BC emissions and share best practices and forecasting exercises. By improving our pooling of information, we can help countries better manage natural disasters caused by melting glaciers.

The loss of glaciers in the mountain ranges of South Asia will have serious consequences for communities and livelihoods across the region, both through immediate threats to water resources and more broadly through greater exposure to climate-induced stress. This report should be of great value to all those concerned with the development of the region in the coming decades.

Hartwig Schafer
Vice President, South Asia Region
World Bank

Acknowledgments

This book was prepared by a team led by Muthukumara Mani (lead economist) under the guidance of Hans Timmer (chief economist) and Hartwig Schafer (vice president) of the South Asia Region of the World Bank. The core team members included Thomas Mosier, Kiran Pandey, Gary Kleiman, Jorge Escurra, Kavita MacLeod, Michael Norton, and Takahiro Yamada. The team was supported in the climate and black carbon modeling work by a team of eminent researchers, including Rebecca Adams-Selin, Matthew J. Alvarado, Richard Armstrong, Christopher Brodowski, Mary Jo Brodzik, Judith Curry, Gregory Faluvegi, Eric Hunt, Rajesh Kumar, Chantelle R. Lonsdale, Drew Shindell, Violeta Toma, Peter Webster, and Ekbordin Winijkul.

The team greatly benefited from insightful comments and guidance from internal peer reviewers—Sameer Akbar, Nagaraja Rao Harshadeep, Claudia Sadoff, Ernesto Sanchez-Triana, and Elena Strukova. The team is also grateful to other colleagues from the World Bank for their thoughtful comments and suggestions at various stages of the book's development, including, most notably, Junaid Ahmad, Najy Benhassine, Cecile Fruman, Faris H. Hadad-Zervos, Abhas Jha, Elena Karaban, Henry G. R. Kerali, Kseniya Lvovsky, William Rex, Lynne Sherburne-Benz, Shaffiq Somani, Mercy Tembon, Govinda R. Timilsina, and William Young.

The team was ably guided by an external advisory committee, including Piers Forster (University of Leeds), Anil Kulkarni (Divecha Centre, Indian Institute of Science), Arun Shrestha (International Centre for Integrated Mountain Development), and Yangyang Xu (National Center for Atmospheric Research). The team would also like to thank participants at the "Cryosphere, Glacier Melting, and Implications on Mountain Economy in the HKHK Region: Trans-boundary Solutions for Resilient Mountain Society" conference jointly organized by the Nepal Ministry of Forest and Environment, the World Bank, the Centre for Green Economy Development–Nepal (CGED), and the International Centre for Integrated Mountain Development (ICIMOD), held in Kathmandu, Nepal on September 17–18, 2018.

The book could not have been produced without the support of Bruno Bonansea, Yann Doignon, Mary Fisk, Elena Karaban, and Jewel McFadden. Thomas Cohen provided excellent editorial support.

The team gratefully acknowledges timely financial support from the South Asia Water Initiative, a trust fund supported by the United Kingdom's Foreign, Commonwealth and Development Office, Australia's Department of Foreign Affairs, and Norway.

In producing this report, the World Bank emphasizes that climate change initiatives and projects shall respect the sovereignty of the countries involved and notes that findings and conclusions in the report may not reflect the views of individual countries or their acceptance of those views.

About the Editor

Muthukumara Mani is a lead economist in the South Asia Chief Economist's Office. He has more than 25 years of experience leading environmental projects, policy dialogue, analytical work, and capacity-building activities. He has operational experience at both the World Bank and the International Monetary Fund, leading green growth and climate change policy dialogue, including from field office locations. He has delivered several high-impact and policy-relevant knowledge products in the area of sustainable development. Mani has a number of books, policy reports, and peer-reviewed journal articles to his credit.

Abbreviations

APSO	Adaptive Particle Swarm Optimization
BC	black carbon
CCHF	conceptual cryosphere hydrology framework
CDF	cumulative distribution function
CFAN	Climate Forecast Applications Network
CLE	ECLIPSE 5a current legislation
CMIP5	Coupled Model Intercomparison Project, fifth phase
CO_2	carbon dioxide
CSIR	Council of Scientific & Industrial Research
DEM	digital elevation model
ECS	equilibrium climate sensitivity
ENSO	El Niño–Southern Oscillation
eQM	empirical QM
ERAI	ERA-Interim
ETI	enhanced temperature index
GCM	global climate model
GDP	gross domestic product
GFS	Global Forecast System
GHG	greenhouse gas
GISS	Goddard Institute for Space Studies

GLOFs	glacial lake outburst floods
HAR	High Asia Refined
HKHK	Himalaya, Karakoram, and Hindu Kush
ICESat	Ice, Cloud, and Land Elevation Satellite
ICIMOD	International Centre for Integrated Mountain Development
IGM	inverse glacier modeling
IPCC	Intergovernmental Panel on Climate Change
KGE	Kling-Gupta Efficiency
KS	Kolmogorov-Smirnov
MAF	mean annual flow
MALR	moist adiabatic lapse rate
MB	mass balance
MBE	mass balance efficiency
MMM	multimodel mean
NASA	National Aeronautics and Space Administration
OLS	ordinary least squares
OR	orographic relationships
PB	Parajka and Blöschl
PRISM	Parameter Regression on Independent Slopes Model
PW	precipitable water
QM	quantile mapping
Q2Q	quantile-to-quantile
RCP	representative concentration pathway
RMSE	root mean squared error
SCA	snow-covered area
SDI	Silt Density Index
SLE	sea-level equivalent
SRTM	Shuttle Radar Topography Mission
SWAT	Soil Water Assessment Tool
TCR	transient climate response
WD	western disturbance
WRF	Weather Research and Forecasting
WRF-Chem	WRF Chemistry model

Overview

Glaciers around the world are rapidly melting. Human activities are at the root of this phenomenon. Specifically, since the Industrial Revolution, carbon dioxide and other greenhouse gas emissions have raised temperatures, even higher in the poles, and as a result, glaciers are rapidly melting, calving off into the sea and retreating on land. The disappearance of glaciers also means less water for consumption by the population, a lower hydroelectric energy generation capacity, and less water available for irrigation.

The glaciers in the Himalaya, Karakoram, and Hindu Kush (HKHK) mountain ranges are melting faster than the global average ice mass. The HKHK glaciers are retreating at a rate of 0.3 meters per year in the west to 1.0 meter per year in the east. Field, satellite, and weather records confirm that 9 percent of the ice area present in the early 1970s had disappeared by the early 2000s. Scenario studies—for example, Shea et al. (2015)—project that the glacier mass within the Everest region today will decrease 39–52 percent by 2050. The almost 55,000 glaciers in the HKHK mountains store more freshwater than any other region outside of the North and South Poles. They contain estimated ice reserves of 163 cubic kilometers, of which almost 80 percent feeds into three major rivers in South Asia: the Indus, Ganges, and Brahmaputra. The basins of these three rivers are home to 750 million people.

The changes in water supply caused by melting glaciers affect agriculture and human consumption of water as well as the potential for hydropower development and tourism. In the short term, glacier melt contributes to disasters such as flash floods, landslides, soil erosion, and glacial lake outburst floods, with mountain communities especially vulnerable to such disasters. In the short run, the increase in melting water could compensate for the receding groundwater levels downstream. In the long run, however, decreased water flow from glaciers will exacerbate water shortages, endangering the livelihoods of rural communities downstream. The melting and thinning of glaciers may also affect hydropower production, which is a key source of renewable energy for the region. In the short term, increased water flows from melting glaciers could increase the risk of floods that damage hydropower facilities, while in the long term, water flow may become insufficient for operating the facilities. Glacier areas are also important to the national tourism industry. The melting of glaciers threatens both local glacier-related tourism and infrastructure that facilitates tourism in downstream areas. Potential damage to other sectors (infrastructure, hydropower, water supply) will also adversely affect the larger tourist industry.

Climate change is accelerating glacier melt in South Asia by changing the patterns of temperature and precipitation. Apart from glacier melt, global climate change driven by greenhouse gases (GHGs) is also changing the region's water resources as temperature and precipitation change throughout the region. A recent World Bank report estimates that increasing temperatures and changing patterns of monsoon rainfall due to climate change could lower the living standards of half the regional population by 2050 (Mani et al. 2018). Therefore, reversing glacier melt and securing access to water in South Asia are not possible without global measures to mitigate climate change. However, climate change is not the only driver of excessive ice melt in the HKHK mountains.

Recent evidence suggests that deposits of anthropogenic black carbon (BC) are responsible for more than 50 percent of the accelerating glacier and snow melt. Black carbon is a product of incomplete combustion from human activities, such as industrial and vehicular emissions, biomass burning, and forest fires. BC produced and circulated within the region decreases the reflectance of glacier surfaces, increasing glaciers' absorption of solar radiation; it also raises air temperatures, increasing glacier melt. The role of BC has important policy implications because, unlike other GHG emissions, it can be eliminated from the atmosphere if emissions stop. This means that local policies to reduce air pollutants can help to reduce the melting of glaciers.

This book investigates the extent to which BC reduction policies undertaken by South Asian countries may affect glacier formation and melt. Through historic analysis, the book assesses the relative impact that each source of black carbon (for example, diesel engines, brick making, cookstoves fueled by biomass, open fires, kerosene wick lanterns, and agricultural practices) has on snow and glacier dynamics. Through the calculation of emission sensitivities, the book also simulates how BC emissions interact with projected climate scenarios, estimates the extent to which these glacial processes affect water resources in downstream areas of the Indus, Ganges, and Brahmaputra river basins, and presents scenarios until 2040 to align with a reasonable policy-making time horizon.

In a business-as-usual scenario, glacier melt will accelerate. Although uncertainty surrounds these scenarios and there are significant variations within the region, it is clear that the very high level of biomass use and increasing energy demands from coal-fired power plants in South Asia are increasing the amount of BC circulating over the HKHK mountain ranges and threatening to accelerate glacier melt.

The book derives the following policy conclusions:

- *Full implementation of current BC emissions policies in South Asia can reduce BC deposition in the region by 23 percent.* While countries of the region are taking a number of steps to curb BC emissions through enhancing fuel efficiency standards for vehicles, phasing out diesel vehicles and promoting electric vehicles, accelerating the use of liquefied petroleum gas for cooking and through other clean cookstove programs, and upgrading brick kiln technologies, various cost-effective measures are available and have been put in place to curb future BC emissions. Even then, the water released from glacier melt is projected to increase in absolute volume and as a share of total water production in the 2040s in the upstream areas of the Indus, Ganges, and Brahmaputra basins. BC emissions can be reduced by an additional 50 percent by enacting and implementing new policies that are currently economically and technically feasible, reducing glacier melt to current levels.
- *Improving the efficiency of brick kilns could be key to managing BC.* Industry (primarily brick kilns) and residential burning of solid fuel together account for 45–66 percent of regional anthropogenic BC deposition, followed by on-road diesel fuels (7–18 percent) and open burning (less than 3 percent in all seasons). Several factors influence BC emissions from brick making, including the technology used, the fuel source, and how the brick kiln is operated and maintained. Some modest up-front investments could pay off quickly. A recent World Bank report identifies a few cost-effective technology solutions geared toward cleaner brick kiln operations that could be implemented with government incentives (Eil et al. 2020).
- *Cleaner cookstoves and especially cleaner fuels can help to reduce BC.* Several programs in the region have supported cleaner cookstoves, but they have met with only limited success, in some cases due to behavioral aspects or lack of an affordable supply chain. Switching to a cleaner fuel, which in many cases would mean moving from biomass or coal to kerosene or liquefied petroleum gas in the short run and to solar in the long run, holds more promise. In India, the central government and some state governments have launched programs to implement fuel switching among low-income households in urban and rural areas. These efforts have met with some success and could potentially be replicated in other parts of the region.
- *Managing water resources now is key to mitigating the potential impacts of glacier melt.* The current inefficient allocation and use of water aggravates the impacts of melting glaciers on water supply, both upstream and downstream. Improving institutions for basin-based water management and using price signals to influence water use are key elements of more efficient water management.
- *Countries in South Asia need to manage their hydropower and storage resources carefully.* Hydropower from the HKHK mountain systems is a resource for supplying the region's growing energy needs and enhancing economic prosperity through energy trade and security. Hydropower can provide local, national, and global

environmental benefits by reducing the consumption of fuelwood and fossil fuels. Hydropower developers that are still planning and building their projects need to consider the possibility of changing water flows. Increased water flow from glacier melt and more variable precipitation may boost the case for undertaking large water storage projects to stabilize availability over the years.

- *Regional cooperation can be an effective transboundary solution, helping countries in the HKHK region to manage glaciers and related natural assets collaboratively.* The HKHK mountain ranges span 2,400 kilometers across six nations (Afghanistan, Bhutan, China, India, Nepal, and Pakistan). The commonality of BC challenges and uncertainties posed by climate change suggest that adopting joint strategies to counteract predicted changes in snow cover and glaciers can bring many benefits. A first step in regional cooperation is to exchange information about BC emissions, changing water flows, best practices, and forecasting exercises. Improved information sharing can also help countries to manage natural disasters caused by melting glaciers.

This book demonstrates that managing BC emissions in South Asia carries the potential of a "quadruple-win" solution. In addition to reducing BC deposits on glaciers, cleaner cooking and burning can improve local air quality, help to mitigate global climate change (and thus help the region to meet its climate targets), and support the achievement of long-term energy security, especially after countries switch to solar and other clean energy solutions.

Water management policies should not be based on past experience. Current practices lead to a different, more challenging future, so policies must be designed to reverse trends like the melting of glaciers. This book is meant to guide the forward-looking policies needed. Success will require active, agile cooperation between researchers and policy makers so that both groups can continue to learn. To support an open dialogue, the model developed and used in this book is an open-source, state-of-the-art model that is now available for others to use and improve on.

References

Eil, A., J. Li, P. Baral, and E. Saikawa. 2020. *Dirty Stacks, High Stakes: An Overview of Brick Sector in South Asia.* Washington, DC: World Bank.

Mani, M., S. Bandyopadhyay, S. Chonabayashi, A. Markandya, and T. Mosier. 2018. *South Asia's Hotspots: The Impact of Temperature and Precipitation Changes on Living Standards.* South Asia Development Matters. Washington, DC: World Bank.

Shea, J. M., W. W. Immerzeel, P. Wagnon, C. Vincent, and S. Bajracharya. 2015. "Modelling Glacier Change in the Everest Region, Nepal Himalaya." *The Cryosphere* 9 (3): 1105–28. doi: 10.5194/tc-9-1105-2015.

CHAPTER 1

Introduction

Water stress is increasing in South Asia as a result of many factors, including increasing demand resulting from population and economic growth and a highly variable supply of water, both temporally and geographically. The amount of water available per capita is decreasing in many countries of the region, and overall scarcity is increasing. In 2018, an Indian government report concluded that the country was facing the worst water crisis in its history, with about 600 million people at risk of extreme water scarcity and 200,000 people dying each year from a lack of safe water (NITI Aayog 2018). The report projected that, by 2030, India's demand for water would be twice the supply and that the country would lose 6 percent of its gross domestic product (GDP) as a result. Reducing water stress requires understanding the contours of the resource, including the quantity and quality of water and potential changes over time. This effort includes identifying current water sources, assessing the causes of natural changes to them, considering policy actions to arrest the changes, and evaluating the potential impacts on the production and supply of water under different policy regimes.

Glaciers around the world are rapidly melting. Human activities are at the root of this phenomenon. Specifically, since the Industrial Revolution, carbon dioxide and other greenhouse gas emissions have raised temperatures, even higher in the poles, and as a result, glaciers are rapidly melting, calving off into the sea and retreating on land. The disappearance of glaciers also means less water for consumption by the population, a lower hydroelectric energy generation capacity, and less water available for irrigation.

In the Himalaya, Karakoram, and Hindu Kush (HKHK) mountain region of South Asia, glacier and snow melt, along with precipitation, largely determine the availability and flow of water. The HKHK mountain ranges span 2,400 kilometers across six nations (Afghanistan, Bhutan, China, India, Nepal, and Pakistan) and contain 60,000 square kilometers of ice, storing more water than any other region outside of the North and South Poles. Snow and glacier melt from the HKHK play an important role in the timing and magnitude of water availability within the region.

Development has a significant impact on the region's hydrology. Water produced in the HKHK is critical to the livelihoods and economies of local and downstream areas. People residing in the rural areas of these basins derive their livelihoods primarily rural areas of these basins derive their livelihoods primarily from agriculture, which is a vital source of food for people in the region's urban communities. Water from the headwaters also contributes significantly to energy production (Gautam, Timilsina, and Acharya 2013; Mirza et al. 2008).

Climate change is altering the hydrologic cycle in South Asia, with impacts including melting glaciers, loss of seasonal snow, and changes in precipitation that pose significant risks to the stability of water resources in the region. Recent evidence suggests that, in addition to changing temperatures and patterns of precipitation, anthropogenic black carbon (BC) deposition is further accelerating glacier and snow melt in the Himalayas.

This book focuses on the impacts within three of South Asia's major river basins: the Indus, Ganges, and Brahmaputra. Map 1.1 displays the locations of these basins and the glacial areas within them. More than 750 million people live within these three basins, including an estimated 200 million living in the headwater regions (Shrestha et al. 2015). While the origin of water availability (snow melt, glacier melt, and rain) differs within and between each of the study basins, snowfall and glacier melt are important controls on the timing and availability of surface water at many locations, especially in the headwater regions (Immerzeel, van Beek, and Bierkens 2010; Lutz et al. 2014). Each basin has varying levels of dependence on mountain headwaters, demand for water, and ability to absorb changes in the timing of streamflow (for example, through systems such as reservoirs).

Gaps exist in our understanding of the impact of climate change and BC on glacier and snow melt and the implications for downstream hydrology. This book seeks to fill these research gaps by (a) identifying the causes of potential changes to the glacier and snow dynamics in the HKHK mountain ranges, (b) quantifying the extent of changes in the glacier and snow mass, and (c) inferring changes in the availability of water in the Indus, Ganges, and Brahmaputra basins. Changes in regional climates are linked largely to global-scale actions and processes, and the response of glaciers and snow changes must be approached in the context of the need to reduce global greenhouse gas emissions. Nevertheless, local actions are also important for reducing or offsetting the impacts in the short term.

This report is designed mainly for technical experts in client countries, the World Bank, and the development community. The report is also complemented by a policy brief directed toward a broader group of stakeholders, including policy makers and practitioners, academic researchers, journalists in the broader development community, and civil society within South Asia and globally.

MAP 1.1 The Indus (Left), Ganges (Center), and Brahmaputra (Right) Basins in South Asia

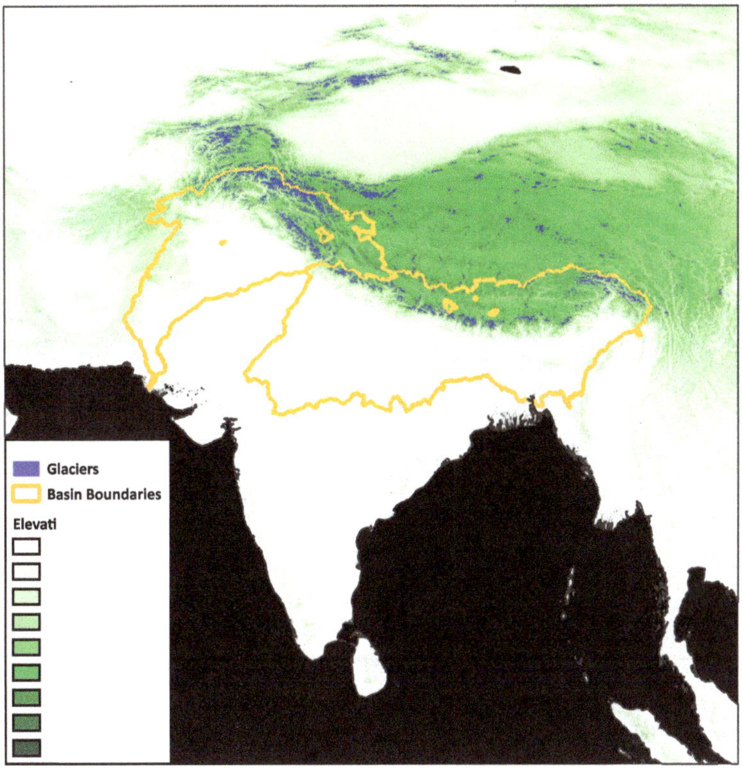

Sources: Based on data from the Food and Agriculture Organization; National Aeronautics and Space Administration, Shuttle Radar Topography Mission; and National Snow and Ice Data Center 2012.

The report is organized as follows. Chapter 2 sets the context by describing the glaciers of the South Asia region, including their economic and environmental benefits and threats that may hasten glacial retreat and melt. Chapter 3 provides detail on the current state of knowledge of two major drivers of glacier melt and retreat in the region: climate change and BC emissions. Chapter 4 details the methodology employed in the research with a focus on closing some of the gaps identified in chapter 3. Chapter 5 looks specifically at the role that BC deposition on snow and ice surfaces plays in altering precipitation patterns and melt rates. Chapter 6 presents the key findings on the potential impacts that aerosol reduction policies of South Asian countries can have on the availability of water resources in the region. Chapter 7 discusses the policy implications of the findings for South Asian countries and the broader global development community. Three appendixes detail the comprehensive analytical work that has informed the research. Appendix A provides information on the selection of a climate model and bias correction. Appendix B details work completed on modeling BC transport and deposition in the South Asia region. Appendix C discusses the methodology for downscaling climate models and describes how global climate models were downscaled to the South Asia region for this book. In addition, appendix D describes the process for calibrating and validating the conceptual cryosphere hydrology framework, and appendix E presents the results for flood risks.

References

Gautam, M. R., G. R. Timilsina, and K. Acharya. 2013. "Climate Change in the Himalayas: Current State of Knowledge." Policy Research Working Paper 6516, World Bank, Washington, DC.

Immerzeel, W. W., L. P. H. van Beek, and M. F. P. Bierkens. 2010. "Climate Change Will Affect the Asian Water Towers." *Science* 328 (5984): 1382–85. doi:10.1126/science.1183188.

Lutz, A. F., W. W. Immerzeel, A. B. Shrestha, and M. F. P. Bierkens. 2014. "Consistent Increase in High Asia's Runoff due to Increasing Glacier Melt and Precipitation." *Nature Climate Change* 4 (7): 587–92. doi:10.1038/nclimate2237.

Mirza, U. K., N. Ahmad, T. Majeed, and K. Harijan. 2008. "Hydropower Use in Pakistan: Past, Present, and Future." *Renewable and Sustainable Energy Reviews* 12 (6): 1641–51.

National Snow and Ice Data Center. 2012. World Glacier Inventory Version 1. Last updated in 2012. Boulder, CO: University of Colorado, National Snow and Ice Data Center.

NITI Aayog. 2018. "Composition Water Management Index: A Tool for Water Management." NITI Aayog, New Delhi.

Shrestha, A. B., N. K. Agrawal, B. Alfthan, S. R. Bajracharya, J. Maréchal, and B. van Oort, eds. 2015. *The Himalayan Climate and Water Atlas: Impact of Climate Change on Water Resources in Five of Asia's Major River Basins.* Kathmandu: International Centre for Integrated Mountain Development (ICIMOD), GRID-Arendal, and CICERO.

CHAPTER 2

Glaciers in South Asia

The glaciers of the Himalaya, Karakoram, and Hindu Kush (HKHK) mountains constitute an important reserve of water in the South Asia region. As noted in chapter 1, the HKHK region contains roughly 60,000 square kilometers of ice, storing more water than any other region except the planet's two polar ice caps and earning it the title of "the third pole." Glaciers are an important part of the hydrologic cycle in South Asia and determine much of the timing and quantity of water flow both locally in the HKHK and in the rivers that originate in the mountains and flow into downstream river basins.

Within the HKHK region, the Indus and Brahmaputra basins have the largest estimated number of glaciers, glaciated area, and ice reserves. Table 2.1 provides details for each river basin in the region on the basin's number of glaciers, glaciated area, estimated ice reserves, and average area per glacier. As shown, glaciers in the HKHK are located in many river basins, but they are concentrated in the Indus, Ganges, and Brahmaputra basins, which together account for nearly 70 percent of the number of glaciers, 74 percent of the total glaciated area, and almost 80 percent of the estimated ice reserves in the HKHK. While many basins are affected by HKHK glaciers, this book focuses on the impacts of glacier change to the Indus, Ganges, and Brahmaputra basins only.

Economic Importance

Glaciers are important to many economic sectors, most notably agriculture. Agriculture provides livelihoods and employment to large numbers of people in the HKHK, and much of it is rain-fed in the mountain areas. For example, 60 percent of Bhutan's population depends mostly on subsistence and rain-fed production; in Nepal, agriculture

TABLE 2.1 Glaciers within the Major Basins of the Himalaya, Karakoram, and Hindu Kush Region

Basin	Number of glaciers	Glaciated area (square kilometers)	Estimated ice reserves (cubic kilometers)	Average area per glacier (square kilometers)
Amu Darya	3,277	2,566	162.6	0.8
Indus	**18,495**	**21,193**	**2,696.1**	**1.2**
Ganges	**7,963**	**9,012**	**793.5**	**1.1**
Brahmaputra	**11,497**	**14,020**	**1,302.6**	**1.2**
Irrawaddy	133	35	1.3	0.3
Salween	2,113	1,352	87.7	0.6
Mekong	482	235	10.7	0.5
Yangtze	1,661	1,660	121.4	1.0
Yellow	189	137	9.2	0.7
Tarim	1,091	2,310	378.6	2.1
Qinghai-Tibetan Interior	7,351	7,535	563.1	1.0
Total	**54,252**	**60,054**	**6,126.9**	**1.1**

Source: ICIMOD 2011.
Note: Totals may not sum due to rounding.

contributes more than 30 percent of gross domestic product (GDP) and employs more than 90 percent of the workforce. In India, agriculture contributes nearly 15 percent of GDP and employs more than 40 percent of the workforce. Overall, around 39 percent of the cropland in South Asia is under irrigation, and irrigated land accounts for 60–80 percent of food production.

The agriculture sector consumes about 90 percent of the water and 20 percent of the total energy used in the region. Studies have shown that changes in temperature and precipitation, monsoon patterns, western disturbances (WDs), and water flow patterns will negatively affect crop productivity in the HKHK and downstream river basins, especially for wheat and maize production. Climate change may reduce crop yields by 30 percent in South Asia. At the same time, demand for cereal will increase, reaching an estimated 476 million tons by 2025, compared to 241 million tons in 2000 (Rasul and Sharma 2016). Glaciers will affect agricultural production to the extent that their melting changes river flows and water supply in both the mountain areas and downstream.

Melting and thinning of glaciers also may affect hydropower production in the HKHK, which is a key source of renewable energy for the region. In the short term, water flows from melting glaciers will increase, creating flood risks that will affect hydropower facilities, while in the long term, water flow will decrease as glaciers fully melt. This is concerning, given the region's high energy needs.

More than 500 million people in South Asia lack access to modern energy. To address this gap, HKHK countries—particularly Bhutan and Nepal—could access large but untapped hydropower potential. These countries will be key for energy exports to meet India's growing demand. In India, 79 percent of total hydropower potential is within the Himalayan region, but only a fraction of that potential has been developed. In Bhutan, only 7 percent of the economically viable production (about 23,800 megawatts) has been developed across five run-of-river hydropower facilities. Hydropower generation is an important export to India and contributes more than 27 percent of Bhutan's revenue and 14 percent of GDP. According to Bhutan's Power System Master Plan 2040, Bhutan has a hydropower potential of nearly 37,000 megawatts. Yet Bhutan had only installed roughly 2,300 megawatts in 2019, or 6.3 percent of the potential; the country plans to install at least 5,000 megawatts by 2030 (Kingdom of Bhutan 2019). In Nepal, 90 percent of energy production comes from hydropower. Against an economically viable capacity of 43,000 megawatts, Nepal has an installed capacity of just 753 megawatts. Nepal needs to harness additional power capacity, as it is subject to energy shortages and blackouts; the country also stands to benefit from revenues associated with energy exports to India.

Glacier areas are also important to the tourism industry. Glaciers themselves constitute a significant tourist resource for many countries of the HKHK, and their melting threatens both local glacier-related tourism as well as infrastructure that facilitates tourism in downstream areas. For example, tourism in Bhutan contributes about 15 percent of annual GDP. High-altitude trekking is popular, but accelerated snow melt in mountain areas threatens trekking routes by increasing the risk of floods and landslides. Potential damages incurred by other sectors (infrastructure, hydropower, water supply) will adversely affect the larger tourist industry in Bhutan.

Glacial Change

While glaciers are critical to economies and livelihoods, the volume of glaciers is decreasing both in the HKHK and globally. In all mountain regions where glaciers exist today, the volume of glaciers has shrunk considerably over the past 150 years, with many small glaciers disappearing. According to a recent study, around 25 percent of the global loss in glacier mass was due to human-related activities between 1851 and 2010, but the figure rose to about 69 percent between 1991 and 2010 (Mishra, Kumar, and Singh 2014). As figure 2.1 shows, the worldwide trend has been toward ever-higher rates of ice loss, especially in the period from 1980 through 2010, when glacier sampling was highest.

Observations over the past century confirm that most of the HKHK glaciers are also retreating. Field, satellite, and weather records confirm that 9 percent of the ice area present in the early 1970s had disappeared by the early 2000s. Between 2003 and 2009, Himalayan glaciers lost an estimated 174 gigatons of water each year and retreated at a rate of 0.3 to 1.0 meters per year (the most extreme melting occurring in the east), a rate faster than the global average ice mass melt (Gurung et al. 2017). According to the latest Intergovernmental Panel on Climate Change (IPCC) regional estimates for Central Asia

FIGURE 2.1 Global Cumulative Change in Glacier Mass

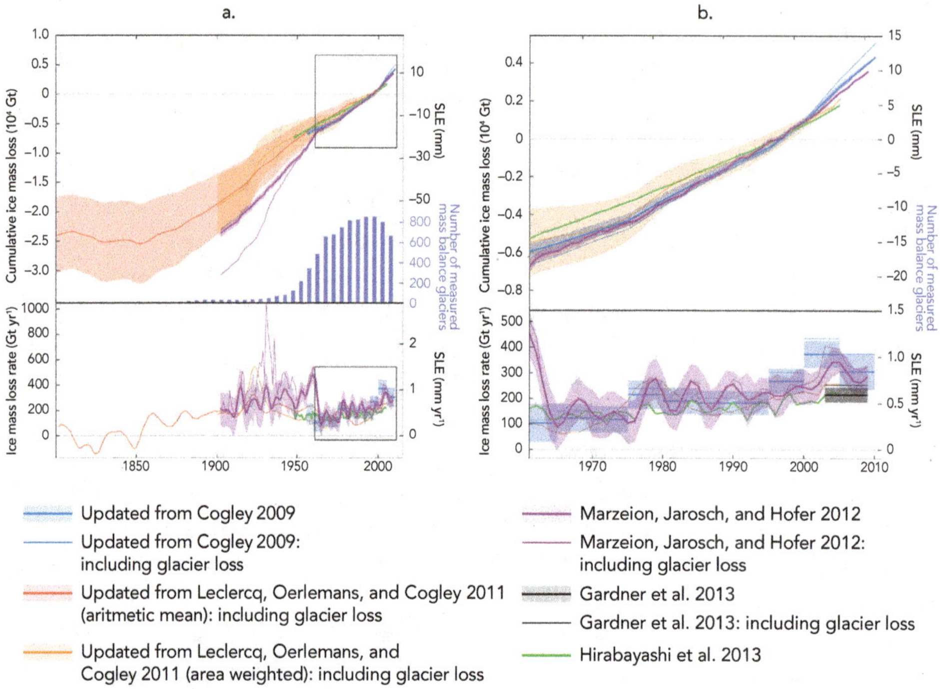

Source: Vaughan et al. 2013.

Note: Global cumulative change in glacier mass for 1801–2010 (whole period) and for 1961–2010 (inset). The cumulative estimates are all set to zero mean over the period 1986–2005. mm = millimeters; SLE = sea-level equivalent.

as well as for the western and eastern glaciers of South Asia, the trends are toward a decline in glacier mass (figure 2.2). Maurer et al. (2019) quantify four decades of ice loss for 650 of the largest glaciers across a 2,000-kilometer transect in the Himalayas. They find similar mass loss rates across subregions and a doubling of the average rate of loss during 2000–2016 relative to the 1975–2000 interval. They suggest that degree-day and energy balance models focused on accurately quantifying glacier responses to air temperature changes (including energy fluxes and associated feedbacks) will provide the most robust estimates of glacier response to future climate scenarios in the Himalayas.

While most glaciers are melting, in some areas—primarily within the Karakoram—glaciers are not retreating; indeed, several glaciers have shown small increases in their mass. Map 2.1 shows the locations of glaciers that have retreated and those that have grown based on satellite-observed differences in elevation (Kääb et al. 2015). These data suggest that glaciers have gained in thickness in the western Kunlun Shan, which is in line with observed in situ measurements of mass balance and change in length (Yao et al. 2012). The data also show a southwest-to-northeast gradient, from considerably negative glacier mass balances in Hindu Kush and Spiti-Lahaul to positive values in the Pamir–Karakoram–western Kunlun Shan region (Kääb et al. 2015). This anomaly seems to be

FIGURE 2.2 Estimated Glacier Mass Budget for Central and South Asia, 1960–2010

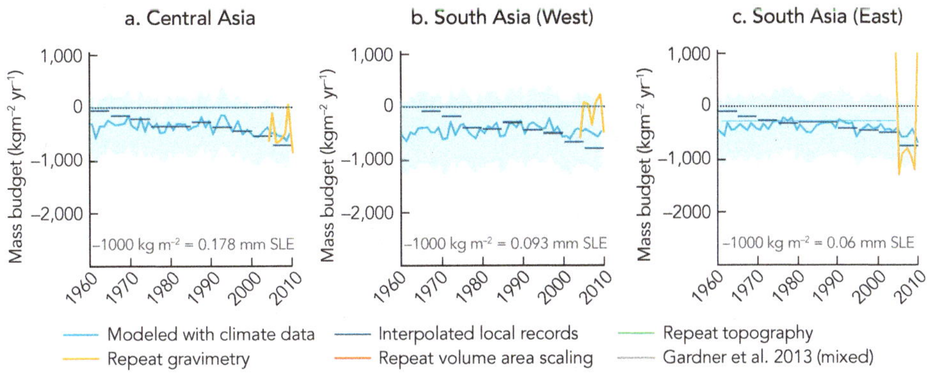

Source: Vaughan et al. 2013.

Note: Regional glacier mass budgets in kilograms per square meter per year (kg/m²/yr) for three of the world's 19 glacierized regions. Mass budget estimates are included only for report domains that cover about 50% or more of the total regional glacier area. Mass budget estimates include 90% confidence envelopes (not available from all studies). Conversions from a specific mass budget in kilograms per square meter (kg/m²) to millimeters (mm) sea-level equivalent (SLE) are given for each region. Most results shown are calculated using a single method, while some merge multiple methods; those from Gardner et al. (2013) are reconciled estimates for 2003–09 obtained by selecting the most reliable results of different observation methods, after region-by-region reanalysis and comparison.

MAP 2.1 Changes Observed in Glacial Elevation in the Himalaya, Karakoram, and Hindu Kush Region, 2003–08

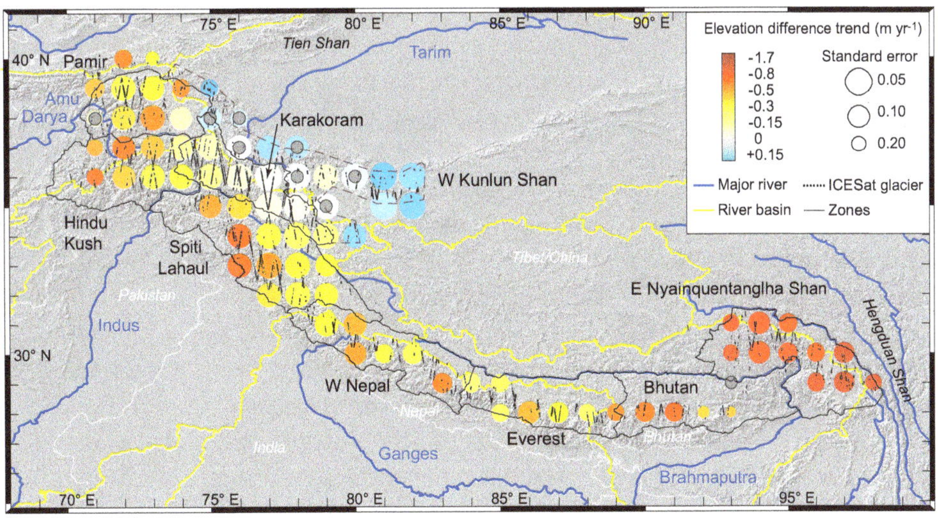

Source: Kääb et al. 2015.

the result of larger-scale meteorological or climatic processes rather than peculiarities of the Karakoram topography or glaciers. Similarly, Salerno et al. (2015) suggest that around Koshi basin, the negative mass balances of glaciers observed in this region can be more ascribed to a decrease in accumulation (snowfall) than to an increase in surface melting.

Implications of Glacial Change

Glacier melt has important implications for the economic sectors that glacier-fed rivers support, particularly in regions where glacier runoff is integral to multiple human activities, such as farming, energy production, supply and quality of water, and tourism. The scale of potential changes facing downstream communities is immense. As noted, the seasonally predictable flow regimes of glacier-fed rivers provide water for hydropower, human consumption, irrigation for agriculture, and tourism. The expected shifts of the hydrologic regime due to glacier and snow melt will have a marked effect on these ecosystem services, with significant economic ramifications (Milner et al. 2017). In the short term, increased water availability is likely to result from continued glacier melt processes. In the long term, however, water resources may decline as glaciers disappear. These future decreases in glacier melt runoff could lead to competition for water resources and worsen water scarcity in the region. Heightened water scarcity would have negative potential consequences for the many environmental, social, and economic sectors that depend on water from the HKHK region.

Accelerated glacier melting in the face of climate change is also expected to exacerbate various water-induced natural hazards, such as glacial lake outburst floods (GLOFs) and general flooding, with subsequent impacts on people and assets at risk. Glacial melt contributes to disasters, such as flash floods, landslides, soil erosion, and GLOFs, which occur when glaciers melt and water collects behind a glacier's terminal moraine (a natural dam of rubble and ice). As these lakes grow and water pressure builds, the moraine can burst, threatening downstream communities and infrastructure. GLOFs have a negative impact on human settlements, infrastructure, hydropower, tourism, agriculture, and other sectors. Roughly 20 glacial lakes in Nepal are currently considered GLOF risks (World Bank 2015). Veh et al. (2018) have identified 19 previously unreported GLOFs in an area covering 57 percent of the HKHK, raising the total count of known GLOFs by 172 percent, based on inventories featuring 11 GLOFs in the region since 1989. Recent research has pointed to an average annual frequency of 1.3 GLOFs since 1980, a figure that has not increased despite the rise in the number of glacial lakes (Veh et al. 2018).

Damages from GLOF events are significant. ICIMOD (2011) estimates the damages from potential GLOFs for three major glacier lakes in Nepal as ranging from US$2.2 billion to US$8.98 billion when hydropower project proposals at the lakes are considered in addition to other damages. Another study in Nepal (Shrestha and Aryal 2010) estimates the total value of properties exposed to potential GLOFs at between US$159 million and US$197 million.

Glacier melt might worsen the risks of flooding in South Asia. Although direct links between glacier melt and the increased number and severity of flooding events need more investigation, it is highly probable that glacier melt will worsen flood risk through GLOFs or rapidly rising water levels from avalanches of ice and short-term heavy rainfall in the monsoon season. Rapid population growth in the region has increased the exposure to flood risk.

Drivers of Glacial Change in South Asia

There are many drivers of change to sustainability in the HKHK mountain range. A recent report notes land use and land change, demographic change, pollution, overexploitation of natural resources, and climate change as challenges to the sustainable availability and use of resources in the HKHK (ICIMOD 2019). These ongoing, dynamic processes contribute to glacial change in varying degrees. While all of these processes are necessarily important and interact in fundamental ways, this book focuses on the impacts of climate change and aerosols—specifically black carbon (BC)—which are key factors in glacial change and melt.

As noted, the scientific evidence shows that Himalayan glaciers are thinning, and human-induced climate change and heavy deposition of BC content play a significant role in this complex phenomenon (Flanner et al. 2009; IPCC 2007; Nair et al. 2013; Ramanathan and Carmichael 2008). Shea et al. (2015) project that glacier mass within the Everest region will decrease by 39 percent by 2050 under representative concentration pathway (RCP) 4.5 and 52 percent under RCP 8.5 relative to the present day.

In addition to the impact of global climate change on precipitation, locally produced BC is a significant factor in glacier melting and retreat. BC is a product of incomplete combustion and is emitted from industry, motor vehicles, power plants, biofuel burning for home use (cooking fires), open biomass burning of forests and crops (Kumar et al. 2015), and other sources. BC circulates through the region and affects the melting of fresh snow through direct radiative interaction with solar and terrestrial radiation via scattering and absorption of sun rays. It modifies the microphysical properties of clouds—both indirectly and semidirectly—and reduces the albedo of snow (Flanner et al. 2007; IPCC 2007; Warren and Wiscombe 1980). The patterns of energy use in the South Asia region, with a very high level of biomass use and increasing demands for energy from coal-fired power plants, are increasing the amount of BC circulating in the atmosphere and over the HKHK mountain ranges. One study on BC emissions from India estimated that total BC emissions were 835.5 gigagrams in 1991 and 1,243.78 gigagrams in 2001, an increase of 49 percent (Sahu, Beig, and Sharma 2008). Notably, coal contributed more than 50 percent of total BC emissions in that study.

The following sections explain in more detail the two main drivers of glacial change: climate change (temperature and precipitation) and BC.

CLIMATE CHANGE

This section details both historic climate trends and projected future climate change scenarios for temperature and precipitation, which are major drivers of glacial change in the South Asia region. Historic climate data in the nearer term are available from in situ monitoring infrastructure and satellites; scientists can obtain data on historic climate over a longer historic period by studying the physical environment—for example, ice cores, tree rings, and sea floor sediment. Historic climate data inform the current understanding of climate change that is under way and provide input to develop and validate climate change models that project probable temperature and precipitation values into the future. Box 2.1 provides additional detail on global climate change modeling.

BOX 2.1 Global Climate Change Modeling

Two essential elements of global climate change modeling are global climate models (GCMs) and representative concentration pathways (RCPs).

GCMs divide the earth into grid cells and complete equations representing physical and thermodynamic processes for each cell. GCMs model the complex interactions between the earth's ocean, atmosphere, and land. More recent earth simulation models have added biogeochemical cycles and carbon. These models use various scenarios to project future climate and climate change. Currently, the scenarios that the global scientific community uses under the Intergovernmental Panel on Climate (IPCC) are referred to as RCPs.

The IPCC's Fifth Assessment Report uses the term RCP to define each future climate scenario (IPCC 2013). Where previous IPCC scenarios started with alternative future socioeconomic scenarios, RCPs start with scenarios of greenhouse gas (GHG) and aerosol concentrations in the atmosphere. Specifically, climate scientists have defined four possible RCP scenarios that they use as consistent inputs for calculating climate in the future (table B2.1.1). Each scenario is based on a plausible future pathway regarding global emissions of greenhouse gases (figure B2.1.1). The scenarios (RCPs) specify the amount of radiative forcing in 2100 relative to 1750 (figure B2.1.2).

FIGURE B2.1.1 Climate Modeling Timeline

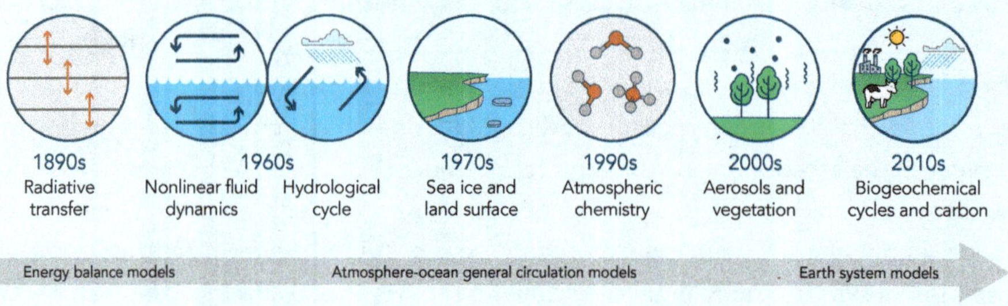

Source: US Global Change Research Program 2017.

TABLE B2.1.1 Representative Concentration Pathway Scenarios

Representative Concentration Pathway (RCP)	Forcing compared to 1750 (watts per square meter)	Climate policy associated with scenario	Carbon dioxide (CO_2) equivalent (parts per million)	Projected global average temperature increase, 1986–2005 (°C)
2.6	2.6	Mitigation	475	1.0
4.5	4.5	Stabilization	630	1.8
6.0	6.0	Stabilization	800	2.2
8.5	8.5	None	1,313	3.7

Source: IPCC 2013, box SPM.1.

(box continues next page)

BOX 2.1 Global Climate Change Modeling *(continued)*

FIGURE B2.1.2 Radiative Forcing Estimates in 2011 Relative to 1750 and Aggregated Uncertainties for the Main Drivers of Climate Change

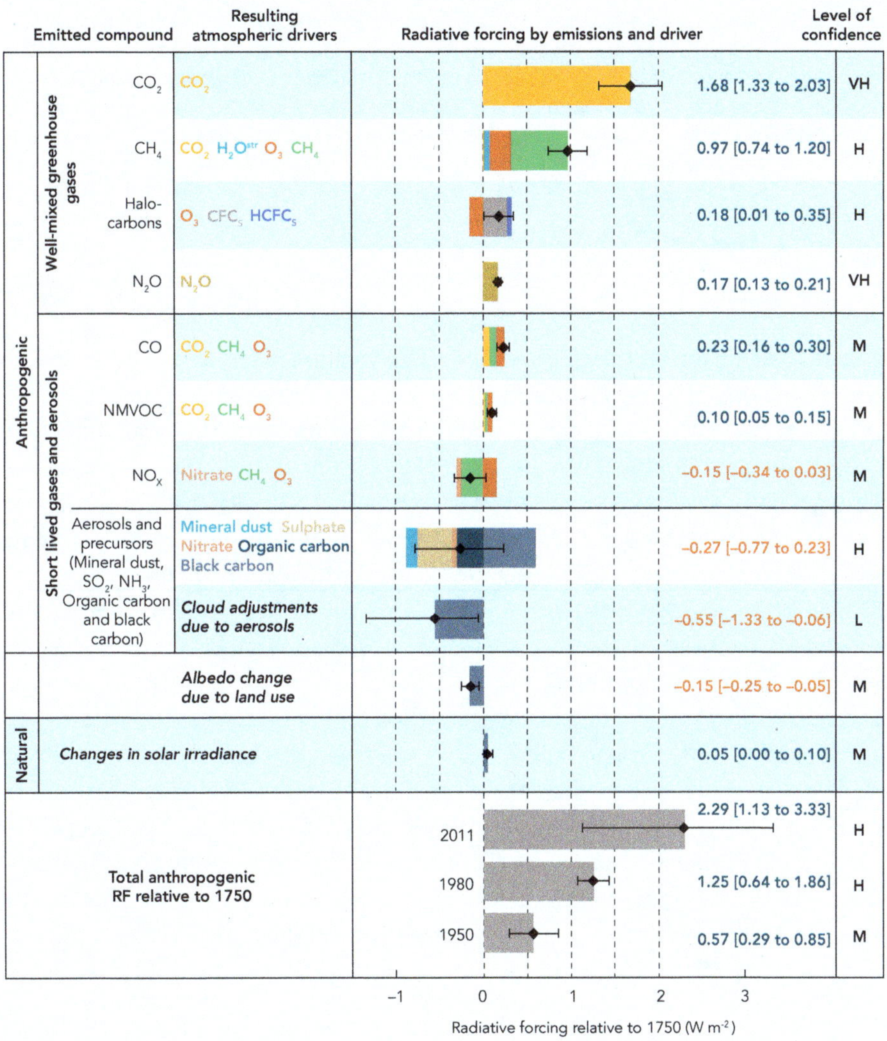

Source: IPCC 2013, box SPM.1.

Note: Values are global average radiative forcing (RF14), partitioned according to the emitted compounds or processes that result in a combination of drivers. The best estimates of the net radiative forcing (RF) are shown as black diamonds with corresponding uncertainty intervals; the numerical values are provided on the right of the figure, together with the confidence level in the net forcing (VH = very high, H = high, M = medium, L = low). Albedo forcing due to black carbon (BC) on snow and ice is included in the BC aerosol bar. Small forcings due to contrails (0.05 W/m^2, including contrail-induced cirrus), and hydrofluorocarbons (HFCs), perfluorocarbons (PFCs), and sulfur hexafluoride (SF6) (total 0.03 W/m^2) are not shown. W/m^2 = watts per square meter. Concentration-based RFs for gases can be obtained by summing the like-colored bars. Volcanic forcing is not included, as its episodic nature makes it difficult to compare to other forcing mechanisms. Total anthropogenic radiative forcing is provided for three years relative to 1750.

(box continues next page)

BOX 2.1 Global Climate Change Modeling *(continued)*

GHGs and aerosols emitted by anthropogenic activity have varying radiative forcing properties (figure B2.1.2). Positive values indicate an overall warming effect; negative values indicate a cooling effect. Various aerosols and their precursors have forcing impacts that vary from cooling to warming; BC is noticeably associated with an overall warming impact. When assessing radiative forcing estimates, it is important to note that the radiative forcing established by specific emitted compounds can be made larger or smaller by feedback loops initiated by other climate drivers. Higher GHG concentrations in the atmosphere increase evaporation and therefore water vapor; increased water vapor in the atmosphere can, in turn, magnify GHG forcing.

TEMPERATURE

The South Asia region is already experiencing an increase in annual mean surface temperature of about 1.5°C relative to the preindustrial average. Observational evidence indicates that the surface temperatures on the Tibetan plateau have increased by about 1.8°C over the past 50 years (Wang et al. 2008). Temperature data in the HKHK overwhelmingly show a warming trend, albeit at different rates in different periods, depending on the regions and seasons. The overall warming in the region over the past century is equivalent to an increase of between 0.9°C and 1.6°C, with the majority of warming occurring in the past 50 years (Gautam, Timilsina, and Acharya 2013). Map 2.2 shows the change in temperature between 1950 and 2010 in South Asia; as shown, temperatures have risen in much of the region.

Longer-term warming trends in parts of South Asia will continue as a result of anthropogenic climate change (Hijioka et al. 2014; see also box 2.2). Averaged across the South Asia region, the median projection for 49 global climate models of the fifth phase of the Coupled Model Intercomparison Project (CMIP5) in RCPs 4.5 and 8.5 is an increase in daily average temperature of 2.1°C and 2.9°C, respectively, by 2046–75 relative to the current average. The projected rise in temperature is slightly higher in winter than in summer and greater in the high-altitude areas in the north.

PRECIPITATION

Precipitation patterns are also changing, although future projections are variable. Gautam, Timilsina, and Acharya (2013) draw tentative conclusions based on their review of the relevant literature: (a) monsoon and annual precipitation is increasing in Jammu and Kashmir, but precipitation is decreasing in the western Indian Himalayas; (b) winter precipitation is decreasing in the western Indian Himalayas, but increasing in the upper Indus basin (Pakistan); (c) no spatially coherent trend of changes in precipitation is evident in Bhutan and Nepal; and (d) annual, winter, and spring precipitation is increasing in the Yarlung Zangbo river basin in the Chinese Himalayas.

MAP 2.2 Historic Temperature Change in South Asia, 1950–2010

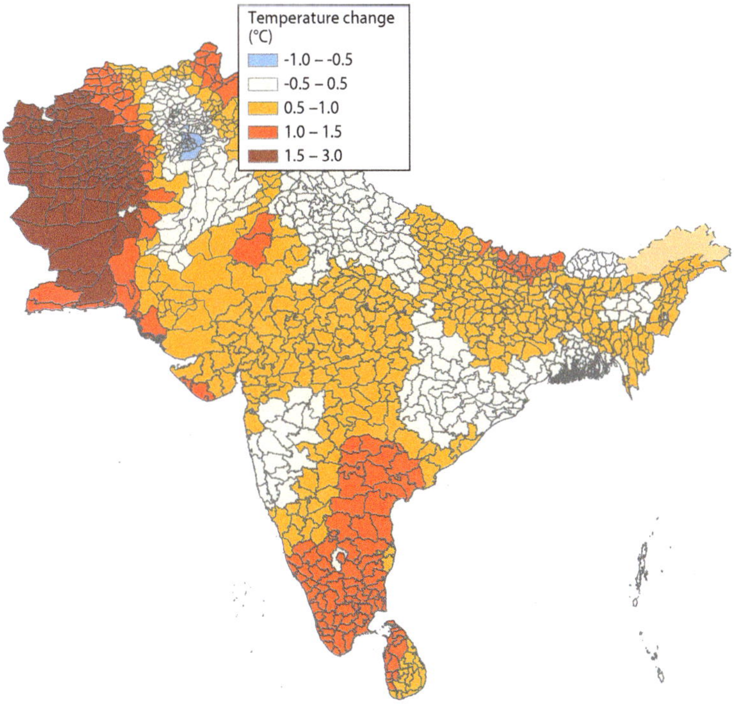

Source: Mani et al. 2018.

BOX 2.2 On Thin Ice: How Cutting Pollution Can Slow Warming and Save Lives

The World Bank and the International Cryosphere Climate Initiative (ICCI) have evaluated the disproportionate impact of climate change on areas of snow and ice known as the cryosphere, with serious implications for human development and environments across the globe (World Bank and ICCI 2013). Their report documents the high and historically unprecedented rate of warming in the cryosphere, which has the potential to trigger disastrous feedback mechanisms in the global climate system. For example, the loss of albedo from sea ice and snow cover and the loss of permafrost may lead to greater carbon fluxes in the atmosphere, particularly where emissions occur as methane.

While the potential implications of a warming climate are alarming, actions to mitigate emissions of short-lived climate pollutants—specifically black carbon and methane—over the next two decades can slow these changes, while benefiting human communities. The mitigation measures modeled for World Bank and ICCI (2013) would deliver multiple health, crop, and ecosystem benefits and would decrease the risks to development from changes in water resources, including flooding and other unforeseen impacts or climate feedbacks. Specifically, in the Himalayas, reducing black carbon emissions from cookstoves, diesel engines, and open burning would have the greatest impact and could significantly reduce radiative forcing and help to maintain a greater portion of Himalayan glacier systems. More detailed modeling at a higher spatial resolution is needed to expand on the work already completed.

Precipitation—specifically the region's monsoon processes—is another important factor in South Asia's hydrology. Together with the Tibetan plateau, the HKHK region exerts great influence on the powerful East Asian and Indian monsoon systems (hereafter, the South Asian monsoon) (maps 2.3 and 2.4), producing a very high climatic gradient across the region. The region's climate ranges from tropical at the base of the foothills to permanent ice and snow at the highest elevations. During the late spring and early summer, the surface of the Tibetan plateau heats up quickly and serves as an elevated heat source, drawing warm, moist air from the Indian Ocean toward the Himalayas and the plateau region. As the monsoon flow transports moisture from the Arabian Sea to the Indian subcontinent, it spurs heavy monsoon rain over the Indo-Gangetic plain and the Bay of Bengal (NRC 2012). The majority of this precipitation occurs from June to September and declines in strength from east to west along the Himalayas.

MAP 2.3 Direction of the Western and Eastern Arms of the Monsoon in India

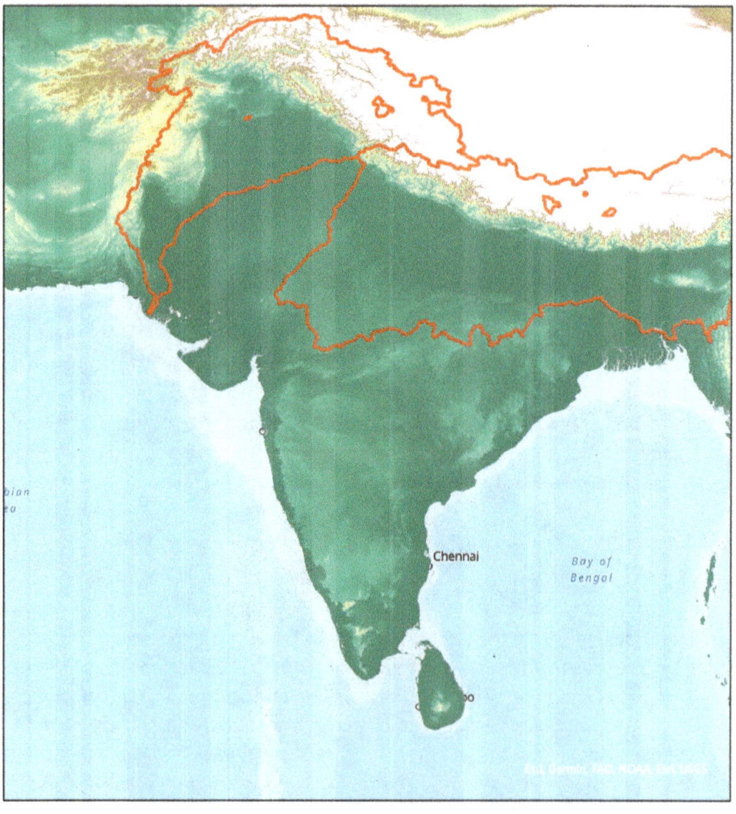

Source: Original map for this publication.

MAP 2.4 Average Annual Monsoon Precipitation in South Asia, 1981–2010

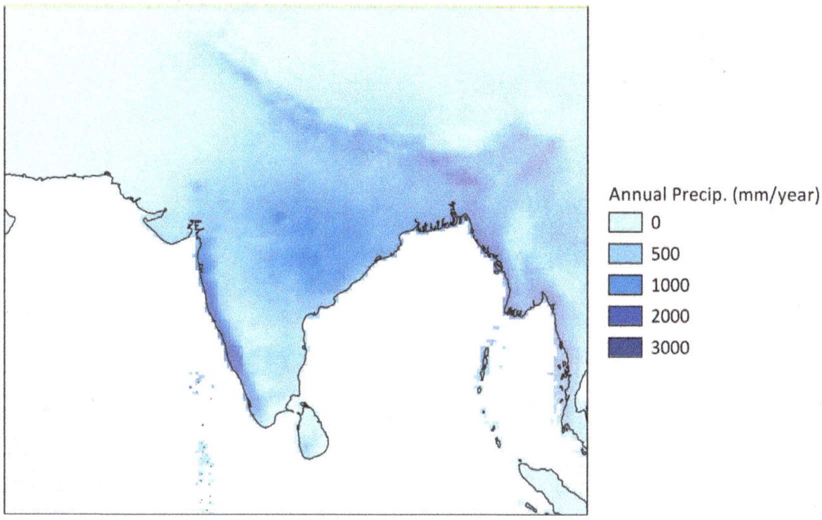

Source: Funk et al., 2015.

The precipitation patterns in the HKHK are characterized by high-level westerly and southwesterly flows, causing precipitation that is distributed more equally over the year, including significant wintertime snowfall in the Karakoram. Up to two-thirds of the annual high-altitude precipitation in the Karakoram occurs in the winter months (Lutz, Immerzeel, and Kraaijenbrink 2014). These westerly driven precipitation patterns tend to weaken as they move from west to east. During the winter, the low-level monsoon flow also reverses to northeasterly, with prevailing large-scale subsidence and relatively dry conditions over India (NRC 2012).

As shown in figure 2.3, overall monsoon precipitation will potentially increase over time, but with significant uncertainty (Harris et al. 2014). The upper reaches of the Ganges and Brahmaputra basins are most affected by monsoon precipitation on both a seasonal and an annual basis, whereas the Indus basin is most affected by year-round (but mostly winter) precipitation from westerly-steering flows.

A number of other weather patterns influence winter precipitation over the western Himalayas, including WDs. These are synoptic-scale cyclonic weather systems advected over Pakistan and northern India by the subtropical westerly jet stream. There, they are responsible for most of the winter precipitation, which is crucial for agriculture of the rabi crop, as well as for more extreme precipitation events, which can lead to local flooding and avalanches. Despite the importance of WDs, there has not yet been an attempt to objectively determine the fate of WDs in climate GCMs (Hunt, Turner, and Shaffrey 2019).

FIGURE 2.3 **Variation in Monsoon Precipitation and Projections in South Asia, 1960–2050**

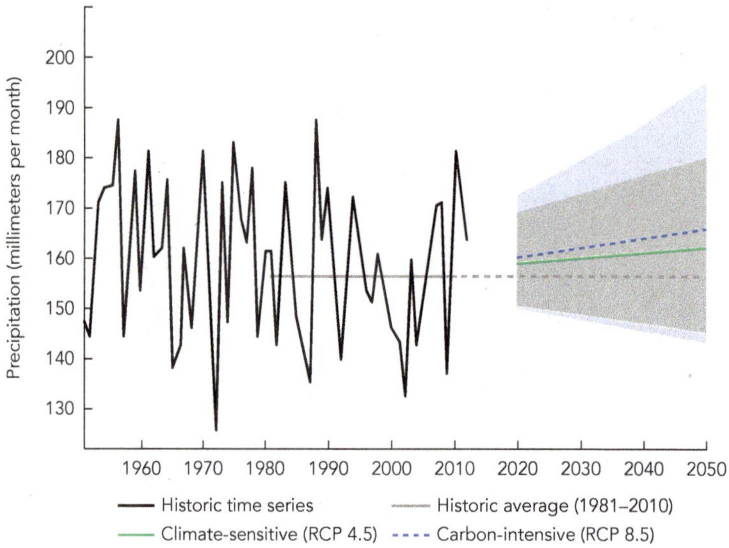

Source: Data from Climate Research Unit TS 2.24 (Harris et al. 2014) and 11 climate models.

Note: The black line indicates yearly monsoon precipitation, the gray line indicates average monsoon precipitation from 1981 through 2010, the dashed purple line indicates the multimodel mean under the carbon-intensive scenario, the green line represents the multimodel mean under the climate-sensitive scenario, and shaded regions indicate the range of results based on 11 climate models for each scenario. RCP = representative concentration pathway.

BLACK CARBON

BC is a short-lived pollutant that is the second-largest contributor to warming the planet behind carbon dioxide (CO_2). Unlike other greenhouse gas emissions, BC is quickly washed out and can be eliminated from the atmosphere if emissions stop. Lower emissions also can improve human health (World Bank and ICCI 2013). However, even small increases in BC emissions and deposition can lead to increased snow melt and sublimation rates, decreased snow albedo and reflectivity, and large net climate radiative forcing. Continuous measurements of BC aerosols carried out at different geographic locations over the Himalayan regions indicate significant BC concentrations in the atmosphere during the premonsoon period. Box 2.3 discusses the mechanisms through which aerosols, including BC, affect regional weather and climate.

BC can increase glacial melt in two primary ways: by decreasing surface reflectance and by raising air temperature. First, the deposition of BC on the surface of snow and ice increases the absorption of solar radiation by glaciers by decreasing the reflectance of the glacier surface (Flanner et al. 2009; Menon et al. 2010). Second, circulating BC also raises air temperatures before it is deposited. In addition to increasing glacial melt, the impact of BC on precipitation may negatively affect glaciers in the South Asia region.

BOX 2.3 Impact of Aerosols on Regional Weather Patterns and Climate

Aerosols affect weather through a variety of physical processes, including scattering and absorption of solar radiation and cloud condensation. These properties directly and indirectly affect temperature and precipitation (Lohmann et al. 2010; Shindell and Faluvegi 2009; Shindell et al. 2012; Stott 2003). Aerosols also affect climate far from the areas in which they are most highly concentrated through changing atmospheric circulation (Wang 2013). In the tropics and extratropical Northern Hemisphere, aerosols have caused a large portion of observed changes in precipitation (Wang 2015).

Climate models that include both direct and indirect aerosol effects are better able to represent observed interdecadal variations in temperature and precipitation (Wang 2015; Wilcox, Highwood, and Dunstone 2013). For example, climate models that include both direct and indirect aerosol effects confirm that a mid-20th-century hiatus in temperature increases was likely related largely to the effects of increased aerosol concentrations (Wilcox, Highwood, and Dunstone 2013). In the tropics and extratropical Northern Hemisphere, aerosols have caused a large portion of observed changes in precipitation (Wang 2015).

There is growing awareness that aerosols are affecting the South Asian monsoon (Bollasina, Ming, and Ramaswamy 2011; Ramanathan et al. 2005; Salzmann, Weser, and Cherian 2014). During the second half of the 20th century, aerosols reduced monsoon precipitation (Bollasina, Ming, and Ramaswamy 2011; Salzmann, Weser, and Cherian 2014). In contrast, it is predicted that, in the absence of aerosols, monsoon precipitation would have increased during this period (Salzmann, Weser, and Cherian 2014).

Variability in rainfall during the South Asian monsoon is also known to be correlated to El Niño–Southern Oscillation (ENSO) phase (Kripalani and Kulkarni 2012; Rajeevan and McPhaden 2004). Severe droughts in India are typically associated with a positive ENSO phase, although a positive ENSO phase is not always associated with drought conditions (Kumar et al. 2006). It is known that the relationship between ENSO and the South Asian monsoon changes over time and that the spatial pattern of sea-surface temperature matters, but there is significant uncertainty surrounding other factors that may influence the relationship (Kumar et al. 2006; Rajeevan and Pai 2007; Torrence and Webster 1999). For example, the extent to which changes in aerosols may affect the relationship between an ENSO phase and the monsoon is not known.

Greenhouse gases (GHGs) and aerosols affect climate through different pathways, and their combined impacts are often challenging to disentangle. Rising temperatures may increase the frequency of positive ENSO excursions, which may increase the number of droughts (Azad and Rajeevan 2016). At the same time, future aerosol emissions are projected to reduce temperature increases compared to a scenario in which GHGs are emitted but aerosols are reduced (Wang et al. 2016; Xu, Lamarque, and Sanderson 2018). As noted, aerosols may also reduce precipitation during the South Asian monsoon.

The timing and quantity of precipitation have the potential to affect the mass balance of glaciers and snow cover and may have a larger influence on hydrology than glacial melt alone. BC is thus significant in both the retreat of some Himalayan glaciers and the changes in basinwide hydrology (Bond et al. 2013; Ramanathan and Carmichael 2008; World Bank and ICCI 2013). The impact of BC on melting glaciers should not be confused with dust- or debris-covered glaciers. Although still inconclusive, some location-specific studies suggest that debris-covered ice has significantly lower melting rates than clean ice. The focus of this report is on climate change and BC-related glacier melt.

The base of scientific evidence underscores the role of BC in glacial melt in the South Asia region, with 69 percent of glacier loss between 1991 and 2011 attributed to the impact of human-related activities such as industrial and vehicular emissions, biomass burning, and forest fires. According to a study by Nair et al. (2013), the increasing presence of BC particles in the atmosphere has potential implications for the regional climate and hydrologic cycle over South Asia. The BC concentration in the atmosphere has been found to be highest during the premonsoon season over the Himalayas, especially in Nepal and the Eastern Himalayas. The direct and surface albedo radiative forcing caused by BC deposition leads to significant warming over the Himalayas during the premonsoon period and thus accelerates glacial melting. Another study by Nair et al. (2013) estimates that BC causes 50 to 90 percent of Himalayan glacier melting. A recent study by the International Centre for Integrated Mountain Development and Thakuri finds that glaciers smaller than 1 square kilometer are disappearing faster and have experienced a 43 percent decrease in surface area since the 1960s (Thakuri et al. 2013). These debris-covered sections of glaciers have increased by about 17 percent since the 1960s. The edges of the glaciers have also retreated by an average of 400 meters since 1962 (Thakuri et al. 2013). A comprehensive study in India covering 146 glaciers in the Chandra basin in the western Himalayas finds that glaciers as a whole lost 19 percent of the total basin volume during the period from 1984 to 2012, with the loss of volume for small and low-altitude glaciers being as high as 67 percent (Pandey et al. 2016). Although some studies document significant interannual variability of mass balances and relatively slower melt rates on debris-covered glacier tongues over interannual time scale, the overall effects of surface debris cover are uncertain, as many satellite observations suggest similar ice losses relative to clean-ice glaciers over similar or longer periods. Because of the complex monsoon climate in the Himalayas, the albedo effect, due to deposition of anthropogenic BC on snow and ice, and precipitation changes have always been suggested as important drivers.

References

Azad, S., and M. Rajeevan. 2016. "Possible Shift in the ENSO-Indian Monsoon Rainfall Relationship under Future Global Warming." *Scientific Reports* 6: 20145. doi:10.1038/srep20145.

Bollasina, M. A., Y. Ming, and V. Ramaswamy. 2011. "Anthropogenic Aerosols and the Weakening of the South Asian Summer Monsoon." *Science* 334 (6055): 502–05. doi:10.1126/science.1204994.

Bond, T. C., S. J. Doherty, D. W. Fahey, P. M. Forster, T. Berntsen, B. J. DeAngelo, and C. S. Zender. 2013. "Bounding the Role of Black Carbon in the Climate System: A Scientific Assessment; Black Carbon in the Climate System." *Journal of Geophysical Research: Atmospheres* 118 (11): 5380–552. doi:10.1002/jgrd.50171.

Cogley, J. G. 2009. "A More Complete Version of the World Glacier Inventory." *Annals of Glaciology* 50: 20–38.

Flanner, M. G., C. S. Zender, P. G. Hess, N. M. Mahowald, T. H. Painter, V. Ramanathan, and P. J. Rasch. 2009. "Springtime Warming and Reduced Snow Cover from Carbonaceous Particles." *Atmospheric Chemistry and Physics* 9 (7): 2481–97. doi:10.5194/acp-9-2481-2009.

Flanner, M. G., C. S. Zender, J. T. Randerson, and P. J. Rasch. 2007. "Present-Day Climate Forcing and Response from Black Carbon in Snow." *Journal of Geophysical Research: Atmospheres* 112 (D11202). doi:10.1029/2006JD008003.

Funk, Chris, Pete Peterson, Martin Landsfeld, Diego Pedreros, James Verdin, Shraddhanand Shukla, Gregory Husak, James Rowland, Laura Harrison, Andrew Hoell, and Joel Michaelsen. "The Climate Hazards Infrared Precipitation with Stations—A New Environmental Record for Monitoring Extremes." *Scientific Data* 2: 150066. doi:10.1038/sdata.2015.66 2015.

Gardner, A. S., G. Moholdt, J. G. Cogley, B. Wouters, A. A. Arendt, J. Wahr, E. Berthier, et al. 2013. "A Reconciled Estimate of Glacier Contributions to Sea Level Rise: 2003 to 2009." *Science* 340 (6134): 852–57.

Gautam, M. R., G. R. Timilsina, and K. Acharya. 2013. "Climate Change in the Himalayas: Current State of Knowledge." Policy Research Working Paper 6516, World Bank, Washington, DC.

Gurung, D. R., S. B. Maharjan, A. B. Shrestha, M. S. Shrestha, S. R. Bajracharya, and M. Murthy. 2017. "Climate and Topographic Controls on Snow Cover Dynamics in the Hindu Kush Himalaya." *International Journal of Climatology* 37: 3873–82.

Harris, I. P. D. J., P. D. Jones, T. J. Osborn, and D. H. Lister. 2014. "Updated High-Resolution Grids of Monthly Climatic Observations: The CRU TS3.10 Dataset." *International Journal of Climatology* 34 (3): 623–42.

Hijioka, Y., E. Lin, J. J. Pereira, R. T. Corlett, X. Cui, G. E. Insarov, R. D. Lasco, E. Lindgren, and A. Surjan. 2014. "Asia." In *Climate Change 2014: Impacts, Adaptation, and Vulnerability. Part B: Regional Aspects*, 1327–70. Contribution of Working Group II to the Fifth Assessment Report of the Intergovernmental Panel of Climate Change. Cambridge, MA: Cambridge University Press.

Hirabayashi, Y., Y. Zhang, S. Watanabe, S. Koirala, and S. Kanae. 2013. "Projection of Glacier Mass Changes under a High-Emission Climate Scenario Using the Global Glacier Model HYOGA2." *Hydrological Research Letters* 7 (1): 6–11.

Hunt, K. M. R., A. G. Turner, and L. C. Shaffrey. 2019. "Falling Trend of Western Disturbances in Future Climate Simulations." *Journal of Climate* 32 (16): 5037–51.

ICIMOD (International Centre for Integrated Mountain Development). 2011. *Glacial Lakes and Glacial Lake Outburst Floods in Nepal.* Report for the World Bank, Global Facility for Disaster Reduction and Recovery (GFDRR). Kathmandu: ICIMOD. https://lib.icimod.org/record/27755.

ICIMOD (Integrated Centre for Integrated Mountain Development). 2019. *The Hindu Kush Himalaya Assessment: Mountains, Climate Change, Sustainability and People,* edited by Philippus Wester, Arabinda Mishra, Aditi Mukherji, and Arun Bhakta Shrestha. Kathmandu: ICIMOD. https://lib.icimod.org/record/34383.

IPCC (Intergovernmental Panel on Climate Change). 2007. *AR4 Climate Change 2007: Synthesis Report.* Geneva: IPCC. https://www.ipcc.ch/report/ar4/syr/.

IPCC (Intergovernmental Panel on Climate Change). 2013. "Summary for Policymakers." In *Climate Change 2013: The Physical Science Basis; Working Group I Contribution to the Fifth Assessment Report of the Intergovernmental Panel on Climate Change,* edited by T. F. Stocker, D. Qin., G.-K. Plattner, M. Tignor, S. K. Allen, J. Boschung, A. Nauels, Y. Xia, V. Bex, and P. M. Midgley. New York: Cambridge University Press. https://www.ipcc.ch/report/ar5/wg1/.

Kääb, A., D. Treichler, C. Nuth, and E. Berthier. 2015. "Brief Communication: Contending Estimates of 2003–2008 Glacier Mass Balance over the Pamir–Karakoram–Himalaya." *The Cryosphere* 9: 557–64.

Kingdom of Bhutan. 2019. "Project on Power System Master Plan 2040 in Bhutan: Final Report." Ministry of Economic Affairs (MOEA) Department of Hydropower and Power Systems (DHPS), Thimphu. https://openjicareport.jica.go.jp/pdf/12326856_01.pdf.

Kripalani, R. H., and A. Kulkarni. 2012. "Climatic Impact of El Niño/La Niña on the Indian Monsoon: A New Perspective." *Weather* 52 (2): 39–46. doi:10.1002/j.1477-8696.1997.tb06267.x.

Kumar, K. K., B. Rajagopalan, M. Hoerling, G. Bates, and M. Cane. 2006. "Unraveling the Mystery of Indian Monsoon Failure during El Niño." *Science* 314 (5796): 115–19. doi:10.1126/science.1131152.

Kumar, R., M. Barth, V. S. Nair, G. G. Pfister, S. Suresh Babu, S. K. Satheesh, K. K. Moorthy, G. R. Carmichael, Z. Lu, and D. G. Streets. 2015. "Sources of Black Carbon Aerosols in South Asia and Surrounding Regions during the Integrated Campaign for Aerosols, Gases, and Radiation Budget (ICARB)." *Atmospheric Chemistry and Physics* 15 (10): 5415–28. doi:10.5194/acp-15-5415-2015.

Leclercq, P. W., J. Oerlemans, and J. G. Cogley. 2011. "Estimating the Glacier Contribution to Sea-Level Rise for the Period 1800–2005." *Surveys in Geophysics* 32: Art. 519.

Lohmann, U., L. Rotstayn, T. Storelvmo, A. Jones, S. Menon, J. Quaas, A. M. L. Ekman, D. Koch, and R. Ruedy. 2010. "Total Aerosol Effect: Radiative Forcing or Radiative Flux Perturbation?" *Atmospheric Chemistry and Physics* 10 (7): 3235–46. doi:10.5194/acp-10-3235-2010.

Lutz, A. F., W. Immerzeel, and P. Kraaijenbrink. 2014. "Gridded Meteorological Datasets and Hydrological Modelling in the Upper Indus Basin: Final Report." FutureWater Report 130, FutureWater, Wageningen.

Mani, M., S. Bandyopadhyay, S. Chonabayashi, A. Markandya, and T. Mosier. 2018. *South Asia's Hotspots: The Impact of Temperature and Precipitation Changes on Living Standards.* South Asia Development Matters. Washington, DC: World Bank.

Marzeion, B., A. H. Jarosch, and M. Hofer. 2012. "Past and Future Sea-Level Change from the Surface Mass Balance of Glaciers." *Cryosphere* 6: 1295–22.

Maurer, J. M., J. M. Schaefer, S. Rupper, and A. Corley. 2019. "Acceleration of Ice Loss across the Himalayas over the Past 40 Years." *Science Advances* 19 (5): No. 6, eaav7266. doi:10.1126/sciadv.aav7266.

Menon, S., D. Koch, G. Beig, S. Sahu, J. Fasullo, and D. Orlikowski. 2010. "Black Carbon Aerosols and the Third Polar Ice Cap." *Atmospheric Chemistry and Physics* 10: 4559–71.

Milner, A. M., K. Khamis, T. J. Battin, J. E. Brittain, N. E. Barrand, L. Füreder, S. Cauvy-Fraunié, et al. 2017. "Glacier Shrinkage Effects on Downstream Systems." *Proceedings of the National Academy of Sciences* 114 (37): 9770–78. doi:10.1073/pnas.1619807114.

Mishra, R., A. Kumar, and D. Singh. 2014. "Long Term Monitoring of Mass Balance of Hamtah Glacier, Lahaul and Spiti District, Himachal Pradesh." *Geological Survey of India* 147 (pt. 8): 230–31.

Nair, V. S., S. Babu, K. Moorthy, A. K. Sharma, A. Marinoni, and Ajai. 2013. "Black Carbon Aerosols over the Himalayas: Direct and Surface Albedo Forcing." *Tellus B: Chemical and Physical Meteorology* 65 (1): 19738.

NRC (National Research Council). 2012. *Himalayan Glaciers: Climate Change, Water Resources, and Water Security.* Washington, DC: National Academies Press. doi:10.17226/13449.

Pandey, P., S. Nawaz Ali, A. Ramanathan, P. Champati Ray, and G. Venkatarama. 2016. "Regional Representation of Glaciers in Chandra Basin Region, Western Himalaya, India." *Geoscience Frontiers* 8 (4): 841–50. doi:10.1016/j.gsf.2016.06.006.

Rajeevan, M., and M. J. McPhaden. 2004. "Tropical Pacific Upper Ocean Heat Content Variations and Indian Summer Monsoon Rainfall." *Geophysical Research Letters* 31 (18).

Rajeevan, M., and D. S. Pai. 2007. "On the El Niño-Indian Monsoon Predictive Relationships." *Geophysical Research Letters* 34: Art. L04704. doi:10.1029/2006GL028916.

Ramanathan, V., and G. Carmichael. 2008. "Global and Regional Climate Changes due to Black Carbon." *Nature Geoscience* 1 (4): 221–27.

Ramanathan, V., C. Chung, D. Kim, T. Bettge, L. Buja, J. T. Kiehl, W. M. Washington, Q. Fu, D. R. Sikka, and M. Wild. 2005. "Atmospheric Brown Clouds: Impacts on South Asian Climate and Hydrological Cycle." *Proceedings of the National Academy of Sciences of the United States of America* 102 (15): 5326–33. doi:10.1073/pnas.0500656102.

Rasul, G., and B. Sharma. 2016. "The Nexus Approach to Water–Energy–Food Security: An Option for Adaptation to Climate Change." *Climate Policy* 16: 682–702. doi: 10.1080/14693062.2015.1029865.

Sahu, S. K., G. Beig, and C. Sharma. 2008. "Decadal Growth of Black Carbon Emissions in India." *Geophysical Research Letters* 35 (2).

Salerno, F., N. Guyennon, S. Thakuri, G. Viviano, E. Romano, E. Vuillermoz, P. Cristofanelli, et al. 2015: "Weak Precipitation, Warm Winters and Springs Impact Glaciers of South Slopes of Mt. Everest (Central Himalaya) in the Last 2 Decades (1994–2013)." *Cryosphere* 9(3): 1229–47. doi:10.5194/tc-9-1229-2015.

Salzmann, M., H. Weser, and R. Cherian. 2014. "Robust Response of Asian Summer Monsoon to Anthropogenic Aerosols in CMIP5 Models." *Journal of Geophysical Research: Atmospheres* 119 (19): 11321–37. doi:10.1002/2014JD021783.

Shea, J. M., W. W. Immerzeel, P. Wagnon, C. Vincent, and S. Bajracharya. 2015. "Modelling Glacier Change in the Everest Region, Nepal Himalaya." *The Cryosphere* 9 (3): 1105–28. doi:10.5194/tc-9-1105-2015.

Shindell, D., and G. Faluvegi. 2009. "Climate Response to Regional Radiative Forcing during the Twentieth Century." *Nature Geoscience* 2 (4): 294.

Shindell, D. T., A. Voulgarakis, G. Faluvegi, and G. Milly. 2012. "Precipitation Response to Regional Radiative Forcing." *Atmospheric Chemistry and Physics* 12 (15): 6969–82.

Shrestha, A., and R. Aryal. 2010. "Climate Change in Nepal and Its Impact on Himalayan Glaciers." *Regional Environmental Change* 11: 65–77. doi:10.1007/s10113-010-0174-9.

Stott, P. A. 2003. "Attribution of Regional-Scale Temperature Changes to Anthropogenic and Natural Causes." *Geophysical Research Letters* 30 (14).

Thakuri, S., F. Salerno, C. Smiraglia, T. Bolch, C. D'Agata, G. Viviano, and G. Tartari. 2013. "Tracing Glacier Changes since the 1960s on the South Slope of Mt. Everest (Central Southern Himalaya) Using Optical Satellite Imagery." *The Cryosphere Discussions* 7: 5389–432. doi:10.5194/tcd-7-5389-2013.

Torrence, C., and P. J. Webster. 1999. "Interdecadal Changes in the ENSO–Monsoon System." *Journal of Climate* 12 (8): 2679–90. doi:10.1175/1520-0442(1999)012<2679:ICITEM>2.0.CO;2.

US Global Change Research Program. 2017. "Climate Models, Scenarios, and Projections." In *Climate Science Special Report,* ch. 4. Fourth National Climate Assessment (NCA4) vol. 1. Washington, DC: US Global Change Research Program.

Vaughan, D. G., J. C. Comiso, I. Allison, J. Carrasco, G. Kaser, R. Kwok, P. Mote, et al. 2013. "Observations: Cryosphere." In *Climate Change 2013: The Physical Science Basis; Working Group I Contribution to the Fifth Assessment Report of the Intergovernmental Panel on Climate Change,* edited by T. F. Stocker, D. Qin, G.-K. Plattner, M. Tignor, S. K. Allen, J. Boschung, A. Nauels, Y. Xia, V. Bex, and P. M. Midgley. Cambridge, UK: Cambridge University Press.

Veh, G., O. Korup, S. Roessner, and A. Walz. 2018. "How Rare Are Himalayan Glacial Lake Outburst Floods?" In *EGU General Assembly Conference Abstracts,* vol. 20: Art. EGU2018-7699, 7699. Munich: European Geosciences Union.

Wang, B., Q. Bao, B. Hoskins, G. X. Wu, and Y. M. Liu. 2008. "Tibetan Plateau Warming and Precipitation Changes in East Asia." *Geophysical Research Letters* 35 (14). doi:10.1029/2008GL034330.

Wang, C. 2013. "Impact of Anthropogenic Absorbing Aerosols on Clouds and Precipitation: A Review of Recent Progresses." *Atmospheric Research* 122 (March): 237–49. doi:10.1016/j.atmosres.2012.11.005.

Wang, C. 2015. "Anthropogenic Aerosols and the Distribution of Past Large-Scale Precipitation Change." *Geophysical Research Letters* 42 (24): 10876–84. doi:10.1002/2015GL066416.

Wang, Z., L. Lin, M. Yang, and Y. Xu. 2016. "The Effect of Future Reduction in Aerosol Emissions on Climate Extremes in China." *Climate Dynamics* 47 (9-10): 2885–99. doi:10.1007/s00382-016-3003-0.

Warren, S., and W. Wiscombe. 1980. "A Model for the Spectral Albedo of Snow. II: Snow Containing Atmospheric Aerosols." *Journal of the Atmospheric Sciences* 37: 2734–45. doi:10.1175/1520-0469(1980)037<2734:AMFTSA>2.0.CO;2.

Wilcox, L. J., E. J. Highwood, and N. J. Dunstone. 2013. "The Influence of Anthropogenic Aerosol on Multi-Decadal Variations of Historical Global Climate." *Environmental Research Letters* 8 (2): Art. 024033. doi:10.1088/1748-9326/8/2/024033.

World Bank. 2015. "Ganges Strategic Basin Assessment: A Discussion of Regional Opportunities and Risks." World Bank, Washington, DC.

World Bank and ICCI (International Cryosphere Climate Initiative). 2013. *On Thin Ice: How Cutting Pollution Can Slow Warming and Save Lives*. A Joint Report of the World Bank and the ICCI. Washington, DC: World Bank.

Xu, Y., J.-F. Lamarque, and B. M. Sanderson. 2018. "The Importance of Aerosol Scenarios in Projections of Future Heat Extremes." *Climatic Change* 146 (3-4): 393–406. doi:10.1007/s10584-015-1565-1.

Yao, T., L. Thompson, W. Yang, W. Yu, Y. Gao, X. Guo, X. Yang, et al. 2012. "Different Glacier Status with Atmospheric Circulations in Tibetan Plateau and Surroundings." *Nature Climate Change* 2 (9): 663–67. doi:10.1038/nclimate1580.

CHAPTER 3

Glaciers and Hydrology in the South Asia Region

The hydrology of the three South Asian river basins included in this book varies across several important parameters and is affected by glaciers and the timing and quantity of glacial melt. Table 3.1 summarizes key details for each river basin, showing the dependence of the watersheds on upstream or downstream precipitation and of the resident populations on the flow for irrigation. The Indus and Brahmaputra basins have extensive upstream areas (that is, areas above 2,000 meters) and larger glaciated areas, while the Ganges and Brahmaputra are generally wetter than the Indus. The Indus, however, is the most dependent on runoff for irrigation (Immerzeel, van Beek, and Bierkens 2010).

The monsoon season brings most of the year's precipitation to South Asia. Map 3.1 displays the percentage of annual precipitation in the South Asia region by season.

The contributions of snow melt, glacier melt, and precipitation to overall runoff vary across basins. This variation can be seen in the breakdown of contributions to overall runoff in map 3.2 (map 3.2 is accessible in the Nontechnical Summary, appendix A, at https://openknowledge.worldbank.org/handle/10986/35600), which presents results from a modeling study for several basins in the Himalaya, Karakoram, and Hindu Kush (HKHK) region (Lutz et al. 2014). The reduced contribution of precipitation to total runoff (green bars) in the upper Indus relative to the upper Ganges and the upper Brahmaputra indicates the reduced role of monsoon precipitation. The map also demonstrates that glacial melt (blue segments of the bars) is a relatively larger contributor in the upper Indus. As a result of these dynamics and as shown in map 3.2, monsoon precipitation has historically dominated the total runoff in the upper Ganges and the upper Brahmaputra,

TABLE 3.1 Characteristics of River Basins in South Asia

Parameter	Indus	Ganges	Brahmaputra
Total area (square kilometers)	1,005,786	990,316	525,797
Total population (thousands)	209,619	477,937	62,421
Annual basin precipitation (millimeters)	423	1,035	1,071
Upstream area (%)	40	14	68
Glaciated area (%)	2.2	1.0	3.1
Annual upstream precipitation (%)	36	11	40
Annual downstream precipitation (%)	64	89	60
Irrigated area (square kilometers)	144,900	156,300	5,989
Net irrigation water demand (millimeters)	908	716	480

Source: Adapted from Immerzeel, van Beek, and Bierkens 2010.
Note: Upstream refers to the areas above 2,000 meters.

MAP 3.1 Average Percentage of Annual Precipitation in South Asia, by Season, 1981–2000

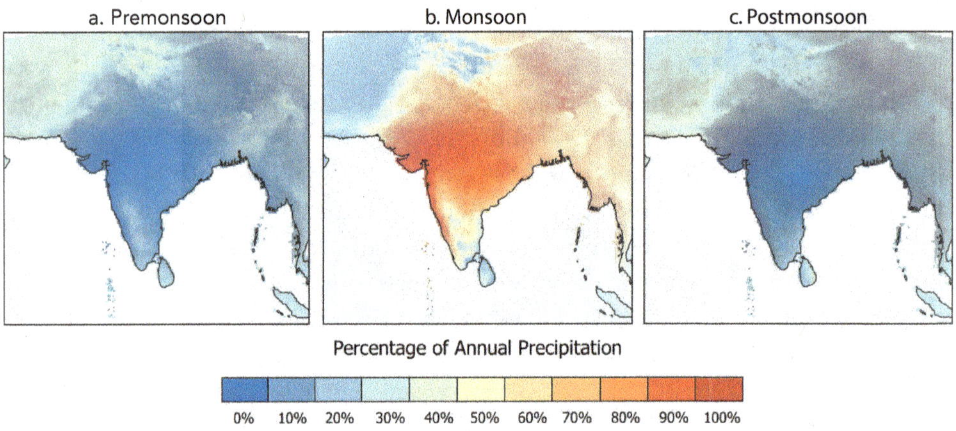

Source: Funk et al., 2015.
Note: Maps are based on average conditions from 1981 to 2000.

occurring mostly during summer months, whereas the high mountains of the Hindu Kush range block the monsoon flow from reaching the Karakoram, where upper-level westerly flow brings snow that appears to have maintained glacial stability (or even growth in some cases).

The sections that follow provide a more detailed examination of the main river basins discussed in this book.

Indus River Basin

A substantial body of research exists on the importance of glaciers in the HKHK to the volume and timing of water in the Indus River and its tributaries as well as the potential impact of climate change to these glaciers. Analyses describing the role of glaciers in the hydrologic regime of these mountains are more limited, however, due in part to the relative inaccessibility of the Himalayan glaciers, which occupy altitudes between roughly 4,000 to 7,000 meters. For this reason, estimates of the potential impacts of glacial retreat are derived, rather than directly observed, using a variety of data sources, including disaggregated low-altitude databases, topography derived from satellite imagery, and simple process models of water and energy exchange in mountain regions.

Principal sources of runoff from the upper Indus basin include winter precipitation and glacier melt. Winter precipitation is a key determinant of the volume of seasonal snow runoff, while summer temperature is a key determinant of the volume of glacier melt. The primary zone of melt water from both winter precipitation and glacier melt is located at between 4,000 to 5,000 meters, due to the combination of maximum-terrain surface area, maximum glacier surface area, and maximum deposition of snow water equivalent occurring there. This altitudinal zone is generally reached by the upward migration of the freezing level during the months of July and August, which are also the time of maximum runoff. More than 60,000 square kilometers of the upper Indus basin are above 5,000 meters—the estimated mean altitude of the summer-season freezing level—and it is generally assumed that significant glacial melt does not occur over most of this upper zone. The glaciers of the region, flowing outward from this zone, have an estimated surface area of approximately 20,000 square kilometers, of which between 7,000 to 8,000 square kilometers are below the summer season freezing level. This 7,000–8,000 square kilometers, in the ablation zone, is the source of the bulk of the annual glacier melt water flowing into the Indus River tributaries.

The principal sources of water in the upper Indus basin are snow and ice melt. According to estimates from a recent World Bank report (World Bank 2013), glacier runoff contributes approximately 16 mean annual flow (MAF) to the total annual flow of the upper Indus basin: 12 MAF from the Karakoram Himalayas, 0.8 MAF from the western Himalayas, and 3 MAF from the Hindu Kush. This flow from glacial runoff represents an estimated 15 percent of the total flow (104 MAF) from the mountain headwaters of the Indus River, with the remaining 85 percent of the total flow representing melt water from the winter snowpack. Stream flow in the upper Indus basin is highly seasonal: approximately 80 percent of the volume of annual stream flow in the tributaries of the upper Indus basin occurs during the summer months of mid-June to mid-September (World Bank 2013).

Ganges-Brahmaputra Basin

Glacial melt in the Ganges-Brahmaputra basin, while likely to increase somewhat, contributes only about 4 percent to basinwide flow. In addition, melting occurs mostly during the high-flow, monsoonal season in the Ganges. In contrast to Europe and North America, where glacier melt contributes to low flows in summer, the Himalayan glaciers melt during the monsoon season, when temperatures are highest and rainfall is heaviest. Thus, while the changing glacier melt will be a fundamental challenge for some melt-dependent mountain communities, it is not a major driver of the Ganges-Brahmaputra basin hydrology.

Within the Ganges basin, the contribution of glaciers to stream flows varies enormously, and glacier melt plays an important role in many glaciated sub-basins. In the Budhi Gandaki basin (Nepal), for example, the contribution of glacier melt to total measured stream flow is about 30 percent, while in the Likhu-Khola basin (Nepal), it is just 2 percent; the average for all of Nepal's rivers is approximately 10 percent. For this reason, while the Ganges system is unlikely to be affected significantly by glacier melt, melting glaciers will have serious local-level impacts. Communities living in glaciated sub-basins that are more dependent on the contribution of glacier melt, for example, could face dramatic changes in the availability of water and a higher incidence of glacial lake outburst floods (GLOFs).

Current State of Research on Himalayan Glacial Dynamics and Melt

Most glacier studies for the HKHK region have focused on glacier dynamics without considering hydrology (for example, Shea et al. 2015) or have investigated the hydrology without modeling dynamic glacier response to climate (Immerzeel, van Beek, and Bierkens 2010). Further, very few studies have examined the impacts of black carbon (BC) on snow and glacier processes. Existing studies with a smaller spatial domain (for example, a single glacier) have tended to use more robust, physically based energy balance models to represent snow and ice accumulation and melt.[1] All previous studies that model large portions of the region have used a type of conceptual formulation called a simple degree-index approach.[2] Two concerns with degree-index models are that they typically do not model the internal energy of snow and ice, and the fitting parameters used tend to change over time (Hock 2005). These deficiencies in simple degree-index models are particularly pronounced when the model is applied for long time scales (projecting the impacts of climate change) or large spatial scales (regional analysis).

Shea et al. (2015) project changes in glaciers in the Everest region caused by climate change but do not model the effect of these changes on the availability of water. They model climate impacts on glaciers using a simple degree-index model to estimate melt and accumulation. In their study, glaciers are allowed to move according to the Weertman (1957) sliding model. The Weertman model does not capture local-scale glacier dynamics, such as the development of pro-glacial lakes and internal glacier deformation, but it is of an appropriate complexity for regional-scale glacier analysis. Shea et al. (2015) also do not incorporate BC into their study. However, they find that glacier mass balances in the Everest region are significantly more sensitive to changes in temperature than changes in precipitation. They also find that the primary way in which higher temperatures increase the rate of glacier mass loss is through expansions of the glacier's ablation zone rather than increases in the melt rate.

The World Bank has also undertaken studies for specific basins within the South Asia region. For example, Yu et al. (2013) investigate the portion of water derived from snow melt versus ice melt in the upper Indus basin. They model the contribution of snow and ice melt through relating a simple degree-index model of snow melt and a regression model of glacier melt to streamflow measurements from several stream gauges in the upper Indus basin. They conclude that projections of precipitation are a major source of uncertainty in understanding the impacts of climate change on streamflow in the region because snow melt is a larger overall portion of the annual water budget than loss of glacier mass. They also find that annual variations in the annual water budget of the upper Indus basin are controlled primarily by differences in the seasonal accumulation of snow and melt; however, in multiple sub-basins within the region, they find that estimated ice melt is equal to snow melt. The study does not include an explicit glacier model, which limits extrapolation of the results to climate change scenarios.

Jeuland et al. (2013) examine the economic impacts of potential changes in water resources in the Ganges basin. Their relatively comprehensive study uses both a reservoir operations model and an economic optimization model to translate potential changes in the water budget into impacts. To understand the effects of climate change projections on streamflow, they also implement the Soil Water Assessment Tool (SWAT) (Arnold et al. 1998), which is a distributed simple degree-index model. As climate inputs to the SWAT model, they use an ensemble of downscaled global climate model (GCM) data for the A2 scenario, which is a climate scenario from the previous generation of GCMs (IPCC 2007). Although the upper portion of the Ganges basin contains significant glacier coverage, Jeuland et al. (2013) do not study glaciers explicitly.

A few hydrologic modeling studies have been conducted at the scale of the Himalayan range (for example, Bookhagen and Burbank 2010; Immerzeel, van Beek, and Bierkens 2010; Lutz et al. 2014). Bookhagen and Burbank (2010) model runoff for 27 major catchments draining the southern Himalayan front, stretching from the Indus in the west to the Brahmaputra River in the east. They use a simple degree-index approach to model the snowpack seasonality but do not include glaciers. Immerzeel, van Beek, and Bierkens (2010) similarly investigate the importance of melt for annual runoff from five of the major river basins—the Brahmaputra, Indus, Ganges, Yangtze, and Yellow—using a simple degree-index model of snow melt and ice melt; they find that melt contributes a greater portion of total annual streamflow in the Indus and Brahmaputra basins than in the other three. They investigate the A1B climate scenario[3] from the Intergovernmental Panel on Climate Change's (IPCC) Fourth Assessment Report (AR4) and include the presence of glaciers, but prescribe future glacier scenarios instead of allowing glaciers to respond dynamically to climate. Lutz et al. (2014) differentiate between debris-free and debris-covered glaciers, but still use a simple degree-index approach and prescribe glacier area scenarios (glacier scenarios are based on Lutz et al. 2013).

Knowledge Gaps

A coupled glacier and hydrologic model for the HKHK remains an analytic gap. While previous studies have either adequately treated glacier dynamics (Shea et al. 2015) or modeled the hydrologic changes without fully coupling the hydrologic and glacier models (Lutz et al. 2014), a coupled glacier and hydrologic model for the entire region has not been attempted. Further, all of the regional studies described in this chapter use a simple degree-index approach; the impact of BC on the availability of surface water has not been studied. Calibration of simple degree-index models varies significantly from grid cell to grid cell and is based on the climatic conditions for which the model is calibrated (Hock 2005; Sicart, Hock, and Six 2008). Jeuland et al. (2013) include the most comprehensive analysis of economic impacts of previous regional studies, but they do not investigate glaciers. No previous glacier hydrology study investigates the economic impacts of water availability.

While several comprehensive inventories have been taken of the size and location of glaciers in the headwater regions, information on the condition of glaciers is limited. For example, the existing inventories assess all perennial snow and ice areas rather than strict glacier cover.[4] For this reason, it is probable that the actual glaciated areas in the inventory are consistently and considerably overestimated: evidence suggests that active glacier ice may be composed of less than 50 percent of the perennial snow and ice areas (Hewitt 2005, 2011; Mayer et al. 2006). In addition, the inventories do not discuss the significant redistribution of snowfall by means of avalanches and wind from off-ice areas to glaciers; this movement likely involves two-thirds or more of all snow that ends up as glacier ice.

The overall impact of climate change and BC on glacier and snow melt remains uncertain. Glacier and snow melt resulting from climate change, BC, and certain characteristics of its components can affect the local and regional physical environment as well as the ecosystem services that people receive from it. These changes can affect economies and communities directly through changes in hydrology, streamflow, ground water recharge, air temperature, disasters, living conditions, infrastructure, and transportation and indirectly through the effects on changing ecosystems and their services (PBS Netherlands Environmental Assessment Agency 2005). The effects of such changes are often nonlinear, interactive, and highly complex; they often cascade from one stress to another, one sector to another, and one place to another. For instance, the melting of glaciers and the resultant GLOFs can trigger other events. Such events may aggravate land degradation, increase variations in the hydrologic regime, and degrade biodiversity, which, in turn, can increase socioeconomic vulnerabilities. Regional glacier hydrology modeling has not accounted for BC.

Prior studies have associated the loss of glacier mass within the region with either global warming or changes in surface reflectivity associated with the deposition of air pollution on glacier surfaces, but not both. For example, Shea et al. (2015) project glacier changes in the Everest region caused by climate change but do not model the effects of air pollution deposition such as BC or the effect of these changes on the availability of water. Shea et al. (2015) project that glacier mass within the Everest region will decrease 39 percent by 2050 under representative concentration pathway (RCP) 4.5 and 52 percent under RCP 8.5 relative to the present day. They assess the climate impacts on glaciers using a simple degree-index model to estimate melt and accumulation; however, their model does not capture local-scale glacier dynamics, such as the development of pro-glacial lakes and internal glacier deformation, or include the effects of BC. Similarly, Kumar et al. (2015) use "tagged tracer" studies of BC to identify the dominant sources of BC in South Asia, but do not analyze the relative impacts of BC deposition. Kopacz et al. (2011) look at the impact of BC deposition on the glacier albedo of the HKHK region, but do not use chemical tagging to determine the primary source regions or sectors associated with the impact. Himalayan glaciers seem to be sensitive to precipitation partly through the albedo feedback on the short-wave radiation balance.

Notes

1. Energy balance models are a class of hydrologic model that purports to capture explicitly all pathways of heat transfer into and out of snow and ice and the resulting energy balance. Since energy balance models represent each snow and ice process explicitly, they are theoretically the most robust and accurate type of model. In practice, all models are incomplete, and each model makes a different set of assumptions about which processes are present and how to represent them.
2. Simple degree-index models are a type of conceptual hydrologic model that is often used in mountain environments because many of the data inputs needed to characterize snow and

ice processes (wind speed and humidity) more explicitly are not widely available or have high uncertainties for these regions. In simple degree-index models, whether snow and ice freezes or melts depends on a threshold temperature (typically a fitting parameter that has a value close to 0°C). When melt occurs, the magnitude of melt is proportional to the air temperature.

3. IPCC's Special Report on Emissions Scenarios (SRES) covers a wide range of the main driving forces of future emissions, from demographic to technological and economic developments. The A1 scenario family develops into three groups that describe alternative directions of technological change in the energy system. The three A1 groups are distinguished by their technological emphasis: fossil intensive (A1FI), nonfossil energy sources (A1T), or a balance across all sources (A1B).
4. The areas inventoried combine permanent snow and glaciers above climatic snow lines or firm limits and active glacier ice below them.

References

Arnold, N., K. Richards, I. Wills, and M. Sharp. 1998. "Initial Results from a Distributed, Physically Based Model of Glacier Hydrology." *Hydrological Processes* 12 (2): 191–219. doi:10.1002/(SICI)1099-1085(199802)12:2<191::AID-HYP571>3.0.CO;2-C.

Bookhagen, B., and D. Burbank. 2010. "Toward a Complete Himalayan Hydrological Budget: Spatiotemporal Distribution of Snow Melt and Rainfall and Their Impact on River Discharge." *Journal of Geophysical Research* 115 (F3): F03019. doi:10.1029/2009JF001426.

Funk, Chris, Pete Peterson, Martin Landsfeld, Diego Pedreros, James Verdin, Shraddhanand Shukla, Gregory Husak, James Rowland, Laura Harrison, Andrew Hoell, and Joel Michaelsen. "The Climate Hazards Infrared Precipitation with Stations—A New Environmental Record for Monitoring Extremes." *Scientific Data* 2: 150066. doi:10.1038/sdata.2015.66 2015.

Hewitt, K. 2005. "The Karakoram Anomaly? Glacier Expansion and the 'Elevation Effect' Karakoram Himalaya." *Mountain Research and Development* 25 (4): 332–40.

Hewitt, K. 2011. "Glacier Change, Concentration, and Elevation Effects in the Karakoram Himalaya, Upper Indus Basin." *Mountain Research and Development* 31 (3): 188–200. doi:10.1659/MRD-JOURNAL-D-11-00020.1.

Hock, R. 2005. "Glacier Melt: A Review of Processes and Their Modeling." *Progress in Physical Geography* 29 (3): 362–91.

Immerzeel, W. W., L. P. H. van Beek, and M. F. P. Bierkens. 2010. "Climate Change Will Affect the Asian Water Towers." *Science* 328 (5984): 1382–85. doi:10.1126/science.1183188.

IPCC (Intergovernmental Panel on Climate Change). 2007. *Climate Change 2007: The Physical Science Basis. Contribution of Working Group to the Fourth Assessment Report of the Intergovernmental Panel on Climate Change.* New York: Cambridge University Press.

Jeuland, M., N. Harshadeep, J. Escurra, D. Blackmore, and C. Sadoff. 2013. "Implications of Climate Change for Water Resources Development in the Ganges Basin." *Water Policy* 15 (S1): 26. doi:10.2166/wp.2013.107.

Kopacz, M., D. L. Mauzerall, J. Wang, E. M. Leibensperger, D. K. Henze, and K. Singh. 2011. "Origin and Radiative Forcing of Black Carbon Transported to the Himalayas and Tibetan Plateau." *Atmospheric Chemistry and Physics* 11 (6): 2837–52. doi:10.5194/acp-11-2837-2011.

Kumar, R., M. Barth, V. S. Nair, G. G. Pfister, S. Suresh Babu, S. K. Satheesh, K. K. Moorthy, G. R. Carmichael, Z. Lu, and D. G. Streets. 2015. "Sources of Black Carbon Aerosols in South Asia and Surrounding Regions during the Integrated Campaign for Aerosols, Gases, and Radiation Budget (ICARB)." *Atmospheric Chemistry and Physics* 15 (10): 5415–28. doi:10.5194/acp-15-5415-2015.

Lutz, A. F., W. W. Immerzeel, A. Gobiet, F. Pellicciotti, and M. F. P. Bierkens. 2013. "Comparison of Climate Change Signals in CMIP3 and CMIP5 Multi-Model Ensembles and Implications for Central Asian Glaciers." *Hydrology and Earth System Sciences* 17 (9): 3661–77. doi:10.5194/hess-17-3661-2013.

Lutz, A. F., W. W. Immerzeel, A. B. Shrestha, and M. F. P. Bierkens. 2014. "Consistent Increase in High Asia's Runoff due to Increasing Glacier Melt and Precipitation." *Nature Climate Change* 4 (7): 587–92. doi:10.1038/nclimate2237.

Mani, M., S. Bandyopadhyay, S. Chonabayashi, A. Markandya, and T. Mosier. 2018. *South Asia's Hotspots: The Impact of Temperature and Precipitation Changes on Living Standards.* South Asia Development Matters. Washington, DC: World Bank.

Mayer, C., A. Lambrecht, M. Belo, C. Smiraglia, and G. Diolaiuti. 2006. "Glaciological Characteristics of the Ablation Zone of Baltoro Glacier, Karakoram, Pakistan." *Annals of Glaciology* 43: 123–31.

PBS Netherlands Environmental Assessment Agency. 2005. *Millennium Ecosystem Assessment: Synthesis Report.* The Hague: PBS Netherlands Environmental Assessment Agency.

Shea, J. M., W. W. Immerzeel, P. Wagnon, C. Vincent, and S. Bajracharya. 2015. "Modelling Glacier Change in the Everest Region, Nepal Himalaya." *The Cryosphere* 9 (3): 1105–28. doi:10.5194/tc-9-1105-2015.

Sicart, J. E., R. Hock, and D. Six. 2008. "Glacier Melt, Air Temperature, and Energy Balance in Different Climates: The Bolivian Tropics, the French Alps, and Northern Sweden." *Journal of Geophysical Research: Atmospheres* 113 (D24). doi:10.1029/2008JD010406.

Weertman, J. 1957. "On the Sliding of Glaciers." *Journal of Glaciology* 3 (21): 33–38. doi:10.3189/S0022143000024709.

World Bank. 2013. *The Indus Basin of Pakistan: The Impacts of Climate Risks on Water and Agriculture.* Washington, DC: World Bank.

Yatagai, A., K. Kamiguchi, O. Arakawa, A. Hamada, N. Yasutomi, and A. Kitoh. 2012. "APHRODITE: Constructing a Long-Term Daily Gridded Precipitation Dataset for Asia Based on a Dense Network of Rain Gauges." *Bulletin of the American Meteorological Society* 93 (9): 1401–15.

Yu, W., T. Yao, S. Kang, J. Pu, W. Yang, T. Gao, H. Zhao, et al. 2013. "Different Region Climate Regimes and Topography Affect the Changes in Area and Mass Balance of Glaciers on the North and South Slopes of the Same Glacierized Massif (the West Nyainqentanglha Range, Tibetan Plateau)." *Journal of Hydrology* 495 (July): 64–73. doi:10.1016/j.jhydrol.2013.04.034.

CHAPTER 4

Data and Methodology

Reliably projecting the hydrologic impacts of regional warming and air pollution—and the black carbon (BC) deposition associated with air pollution—requires a complete snow and glacier hydrology model at high enough spatial and temporal resolution to incorporate realistic scenarios of air pollution and the impacts of its deposition on the reflectivity of snow and ice surfaces. Producing a physically representative snow and glacier hydrology model (often referred to as an energy balance model) requires high-spatial-resolution, low-uncertainty climate products for all variables influencing the snow and glacier energy balance (such as wind speed, humidity, and cloudiness). For regions such as the Himalaya, Karakoram, and Hindu Kush (HKHK), these variables are not well characterized at spatial resolutions sufficient to represent the snow and glacier energy balance accurately and robustly using a fully explicit model. For this reason, this book uses a medium-complexity coupled snow-and-glacier hydrology modeling approach that is more robust than the simple degree-index method used in previous studies but requires fewer data inputs than a full energy balance model. This chapter describes the details of this modeling approach.

Overview

This book examines whether BC policies undertaken by the South Asian countries could have an impact on glaciers and therefore on water resources in the HKHK within the context of a changing global climate. Answering this question requires a methodology that incorporates climate change and BC scenarios into a mountain hydrology model for gauging the overall sensitivity of Himalayan water resources (glaciers and snow) to climate change and aerosols such as BC. The analysis focuses on estimating the incremental impacts that enacting and implementing additional BC reduction polices in South Asian countries would have on the availability of water resources in the 2040s. The time frames were chosen to align with a reasonable time horizon for policy making.

This research builds on a significant body of preceding work. Table 4.1 summarizes previous analyses that have generated key methodologies and inputs used here. Details of these analyses are provided in the appendixes indicated.

This book examines BC mitigation policy in the context of global climate change using a nested modeling approach. In this approach, new global climate model (GCM) simulations were conducted with a low-aerosol (and BC) scenario in the context of representative concentration pathway (RCP) 4.5. This GCM simulation was then used to drive a regional climate model that models BC transport and deposition at a high resolution throughout the HKHK region. The GCM temperature and precipitation outputs and the regional climate model BC output were then input to the conceptual cryosphere hydrology framework (CCHF), which simulates how these input scenarios affect water production, including rain runoff, snowpack formation and melt, and glacier formation and melt across the region. The CCHF includes physical

TABLE 4.1 Previous Analyses Related to the Current Research

Appendix	Analytical component	Document name
A	Climate model selection and bias correction	Mani et al. (2018) review 18 global climate models of the Coupled Model Intercomparison Project (CMIP5), select 11 based on their performance in modeling historic South Asian climate, and use those 11 models as an ensemble to project long-term changes in average temperature and precipitation in the region. Of these 11 climate models, this research used 8 to form ensemble climate projections for the representative concentration pathway (RCP) 4.5 and RCP 4.5 mitigation scenarios.
		Global climate models selected for this research were bias-corrected relative to ERA-Interim to ensure consistent representation of weather patterns across the region for the historic reference period.
B	Black carbon transport and impacts	Black carbon transport was assessed using representative years from the Goddard Institute for Space Studies (GISS) simulation for the historic period and the 2040s.
		Alvarado et al. (2018) examine the differences in aerosol emissions and transport associated with these two scenarios.
C	Downscaling climate	Climate models were downscaled to the South Asia region and compared to other downscaling methods.
		The novel temperature downscaling method derived monthly lapse rates empirically from low-spatial-resolution daily temperature data.

Source: Original compilation for this publication.

Note: ERA-Interim is a climate reanalysis data set, covering the period from 1979 to August 31, 2019. ERA stands for "ECMWF Re-Analysis" and refers to a series of research projects at ECMWF that produced various data sets.

representations of more than 21 physical processes constrained by the topography and initial glacier inventory and state (location, depth, debris cover). The CCHF also uses topography data and initial glacier properties (location, depth, debris cover) as inputs. A simplified schematic of the modeling approach is shown in figure 4.1.

Topography and Glacier Data

The topography data used in the CCHF for the study area were obtained using the National Aeronautics and Space Administration (NASA) Global 30-arcsecond digital elevation model, which is based on the high-quality measurements taken during the Shuttle Radar Topography Mission (SRTMGL30) (Farr et al. 2007; Jarvis et al. 2008). These data provide a 30-arcsecond (approximately 900 meters) spatial grid, which is sufficient to capture spatial heterogeneities important for modeling hydrologic and glaciological processes while being computationally feasible.[1] This is the typical spatial resolution used in regional hydrologic modeling.

Data on the location, depth, and condition of glaciers (glacier debris, depth of debris) used in the CCHF model were compiled from multiple sources. The Randolph Glacier Inventory (RGI 6.0), a global inventory of glacier outlines developed from the Global Land Ice Measurements of Space Initiative, identifies the location of glaciers in the region (RGI Consortium 2017). Data on the thickness and debris cover of glaciers were obtained from Kraaijenbrink et al. (2017).[2]

FIGURE 4.1 Schematic of the Conceptual Cryosphere Hydrologic Framework (CCHF) Modeling Methodology

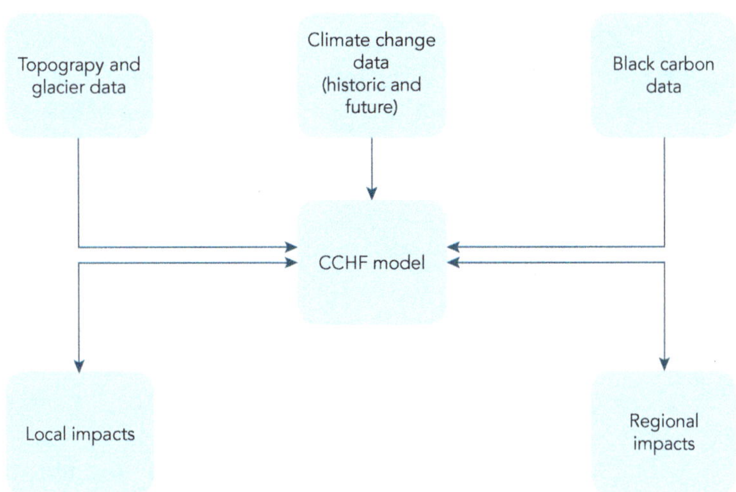

Source: Original figure prepared for this publication.

Climate Data

Greenhouse gases (GHGs) and aerosols (including BC) are both forcing mechanisms in climate change and need to be accounted for separately to distinguish the relative impact of BC on temperature and precipitation. Climate and climate change data provided the necessary inputs on historic and projected temperature and precipitation and allowed for disaggregating the impacts of GHG versus aerosol (BC) on these climate variables. In order to account for the large climatic variations in a topographically diverse mountain region, historic and projected climate data were downscaled to the HKHK region. While downscaling does introduce new uncertainties into the scenarios, it accounts for physically representative processes, such as the impacts of elevation on climate, and therefore is generally beneficial for high-resolution mountain hydrology studies.

The following sections provide a more detailed treatment of the climate data, tools, and approaches used in the research for this book. Box 2.1 in chapter 2 describes global climate models, scenarios, and GHG components.

GLOBAL CLIMATE MODELS

GCMs are the primary tool for projecting how the climate will respond to changes in atmospheric GHG emissions and what regional climate patterns will be in future decades. Climate models seek to represent relevant physical processes and linkages between solar forcing, the atmosphere, land surface, and oceans.

Climate models are imperfect representations of the earth system, and each model makes its own set of assumptions about how to represent specific processes. Climate models are therefore compared along several dimensions. Of particular importance is a climate model's ability to reproduce the impacts of changes in GHG emissions on climate. Two common metrics for quantifying a climate model's sensitivity to changes in atmospheric GHG concentrations are equilibrium climate sensitivity (ECS) and transient climate response (TCR). ECS is the total, long-term change in temperature that would result from a doubling of carbon dioxide (CO_2) in the atmosphere. TCR is the change in temperature averaged over a period of 60–80 years during an experiment in which CO_2 concentrations are increased 1 percent per year (compounded annually).

Climate models also vary in how they represent aerosols. Because aerosols are known to affect the earth's energy budget (Hourdin et al. 2017), the research for this book explicitly sought to understand the impacts of aerosols on climate and BC deposition in the HKHK mountains. Wang (2015) differentiates between climate models that have "Group 1" aerosol representations and "Group 2" aerosol representations. While Group 1 models are the most physically representative with respect to aerosols, Ekman (2013) finds no clear difference between the accuracy of models in Group 1 and Group 2. Further, the uncertainties associated with model parameterizations related to aerosols and cloud microphysics are difficult to deconvolute because climate models are often calibrated so that there is an effective trade-off between the magnitude of the indirect aerosol effect and the sensitivity to CO_2.

CLIMATE MODEL SELECTION

The research for this book leveraged existing work completed on selecting GCMs appropriate for the South Asia region.[3] The ensemble of GCMs in Mani et al. (2018) were assessed further, and 8 of the 11 climate models were selected as an ensemble because they best represent historic climate in South Asia and have publicly available data for the scenarios modeled here. These selected models were used to form ensemble climate projections for the RCP 4.5 (standard) and RCP 4.5 (mitigation) scenarios developed for this research. Details on each of the models with respect to GHG sensitivities and attributes are shown in table 4.2.

TABLE 4.2 Aspects of Climate Modeling

Climate model	Equilibrium climate sensitivity (°C)	Transient climate response (°C)	Aerosol group	Indian monsoon precipitation (climatology / interannual variability)	Model reference
CANESM2	3.7	2.4	1	0.82 / 0.01	Arora et al. (2011)
CCSM4	2.9	1.8	2	0.85 / 0.34	Gent et al. (2011)
CNRM-CM5	3.3	2.1	2	0.85 / 0.25	Voldoire et al. (2012)
GFDL-ESM2M	2.4	1.3	2	0.83 / 0.25	GFDC Global Atmospheric Model Development Team (2004)
GISS-E2-R (p3)	2.1[a]	1.5[a]	1	0.73[a] / 0.38[a]	Schmidt et al. (2014)
IPSL-CM5A-LR	4.1	2.0	2	0.80 / 0.61	Dufresne et al. (2013)
MIROC-ESM-CHEM	4.7*	2.2[b]	1	0.64 / 0.05	Watanabe et al. (2011)
MPI-ESM-MR	n.a.	2.0	n.a.	0.79 / 0.40[c]	Giorgetta et al. (2013)
NorESM1-M	2.8	1.4	1	0.85 / 0.52	Kirkevåg et al. (2013)
Multimodal mean	3.3	1.9	n.a.	0.80 / 0.31	n.a.

Sources: These nine models were selected from Mani et al. (2018) as best representing historic conditions in South Asia. Equilibrium climate sensitivity and transient climate response values are from IPCC (2013a, table 9.5). The aerosol group is from Wang (2015). Monsoon precipitation is from Sperber et al. (2013).

Note: n. a. = not available. CMIP5 = Coupled Model Intercomparison Project, fifth phase. GFDC = Geophysical Fluid Dynamics Laboratory.

a. Values for GISS-E2-R are based on values for the CMIP5 model version, which does not use Physics Package 3. Physics Package 3 includes online aerosol interactions, making it a Group 1 aerosol model.

b. Values for MIROC-ESM-CHEM are based on values for the CMIP5 model version. This version may differ from the "Chem" variant.

c. Values for MPI-ESM-MR are taken to be the values for MPI-ESM-LR because MPI-ESM-MR was not evaluated.

As shown in table 4.2, climate models contain a wide variety of sensitivities to GHG emissions. The nine models for which statistics are provided are those determined to represent historic climate in South Asia most closely (Mani et al. 2018). While these models represent historic climate well, they vary markedly with respect to their ECS and TCR. The ensemble also contains Group 1 and Group 2 models with respect to aerosol representations. These differences point to the imperfect nature of models and suggest that there is no single best practice for constructing a state-of-the-art climate model or an ensemble of climate models.

BIAS CORRECTION

Climate model simulations are often bias corrected to ensure that their representation of climate is consistent with reference data sets. Many types of bias correction exist, and each method makes different sets of assumptions. A central tenant of bias correction methods is that climate models better represent the sensitivity of climate to changes in external forcings than climate corresponding to a specific set of external forcings. Stated more simply, climate models are assumed to represent changes in climate over time more accurately than climate at a specific time.

The selected GCMs (table 4.2) were bias corrected relative to ERA-Interim to ensure that weather patterns across the region are represented consistently for the historic reference period (Dee et al. 2011). ERA-Interim is a reanalysis model based on the European Centre for Medium-Range Weather Forecasts model, which uses available observations to force a physically based weather model implemented at a spatial resolution of 80 kilometers for the entire globe. For this region, ERA-Interim is superior to data sets that are strictly observational in nature because it models weather in remote regions where there are few or no observations. The bias correction method used in this book is quantile mapping, which compares the cumulative distribution function of the simulation output and reference data at each grid cell and then applies this "delta" to the simulation output for both the historic comparison and future projection periods (that is, RCP 4.5 in the 2040s) (Mosier, Hill, and Sharp 2017). This process was conducted for all of the climate models in both the standard RCP 4.5 and RCP 4.5 mitigation scenarios. The historic baseline period used for bias correction is 1986 to 2015. Only 9 of the 11 climate models used in Mani et al. (2018) were assessed because these are the ones for which the necessary daily simulations are publicly available. Additional details on climate model selection and bias correction are provided in appendix A.

Creating the Black Carbon Scenarios

While the RCPs span a wide range of total radiative forcing, they do not cover the full range of emissions, particularly for aerosols (IPCC 2013b). Unlike GHGs, BC is more heterogeneous, and the patterns of BC transport and deposition vary. At the same time, like GHGs, aerosols are known to affect weather through a variety of physical processes, including scattering and absorption of solar radiation and cloud condensation. These properties directly and indirectly affect temperature and precipitation (Lohmann et al. 2010; Shindell and Faluvegi 2009; Shindell et al., 2012; Stott 2003). Aerosols are also known to affect climate far from areas with the highest concentrations by changing atmospheric circulation (Wang 2013).

BC has generally not been modeled at a large scale over long periods of time. In this book, BC transport was assessed using representative years from the Goddard Institute for Space Studies (GISS) simulation for the historic and 2040s periods as input to the Weather Research and Forecasting (WRF) coupled with Chemistry (WRF-Chem) model (Grell et al. 2005) to understand the impacts of regional emissions and transport on the wet and dry deposition of BC in the HKHK region at high spatial resolution. WRF-Chem is a regional climate model that accurately represents atmospheric chemistry and tracks transport of BC from emission source to deposition; it was implemented at a spatial resolution of 12 kilometers for India and the HKHK region. The BC deposition data from WRF-Chem were then used as an input to the CCHF to understand spatial and temporal heterogeneities in BC deposition under the historic and future scenarios across the HKHK region. The scenarios also included a sensitivity analysis of the three distinct phases of the El Niño–Southern Oscillation (ENSO) cycle.

Two sets of experiments were run with the GISS-E2-R climate model (Schmidt et al. 2014) to assess the impacts of aerosols emitted in South Asia for the years 2006 through 2100. The RCP 4.5 simulation was selected as a pathway; RCP 4.5 assumes that much of the world is taking moderate actions to reduce their GHG and BC emissions; it represents a future state where the influence of South Asian sources will be relatively more important. This RCP scenario thus provides a more policy-relevant basis for examining potential policy actions within the nations of South Asia. The first scenario, referred to as RCP 4.5 (standard), follows the RCP 4.5 emissions trajectories for GHG and aerosol emissions.[4] The second scenario, referred to as RCP 4.5 (mitigation), follows the RCP 4.5 emissions trajectories for both GHG and aerosol emissions, except that aerosol emissions in South Asia were modified to equate to ECLIPSE scenario levels (Stohl et al. 2015), as described in the paragraphs that follow.

First, the residential combustion inventory of the ECLIPSE inventory was distributed following the residential combustion inventory of Winijkul, Fierce, and Bond (2016) and Winijkul and Bond (2016). The Winijkul data set provides emissions from residential combustion in 2010 using geographic information system data to allocate population and resource availability (fuelwood, electricity). Access to fuelwood in this inventory was determined using the night light product from NASA satellites to identify access to electricity and other data. Recent emission measurements of different stove technologies, including BC from wick lamps, were also included.

Second, the ratio of the RCP 4.5 pathway to that of the ECLIPSE 5a current legislation (CLE) scenario was used to scale the combined ECLIPSE and Winijkul, Fierce, and Bond (2016) inventories for all scenarios. This allowed the RCP 4.5 pathways to provide the rate of change of total emissions and the ECLIPSE and Winijkul, Fierce, and Bond (2016) data sets to provide data on emissions from each type of source and spatial distribution of emissions.

Two scenarios for future anthropogenic emissions covering the range of minimum and maximum emission reductions were examined (Stohl et al. 2015):

- The *no further control scenario* uses the same assumptions as current legislation until 2015 but assumes that no further legislation is introduced subsequently, even if currently committed. This scenario leads to higher emissions than the CLE scenario for most pollutants and, after using the scaling approach employed in this book, results in higher emissions than the RCP 4.5 scenario as well.
- The *ECLIPSE mitigation scenario* includes all measures with beneficial air quality and climate impact. These measures range from replacing kerosene wick lamps with LED lamps to providing households with cleaner cookstoves. The emission reduction calculations for all pollutants included in the mitigation scenario are consistent with the approach presented in UNEP (2011) and Shindell et al. (2012).

Differences in aerosol emissions and transport associated with these two scenarios are examined in chapter 5 and appendix B (Alvarado et al. 2018).

The runs of the novel GISS model provide a basis for understanding how aerosols (including BC) move between regions (for example, dust travels from the Middle East to Asia) and at a broad scale within the South Asia region. The differences in temperature and precipitation present in the GISS RCP 4.5 (standard) and RCP 4.5 (mitigation) scenarios were then applied to the other GCMs in the ensemble using the same bias correction method described in the preceding section. Results from the GISS modeling exercise and bias correction are presented in appendix B.

Downscaling Climate in the Himalayas

Precipitation and temperature estimates in mountainous regions can vary dramatically over small areas due to high variation in topographical features. For this reason, lower-resolution weather or climate data must be downscaled to a higher resolution to fit certain geographies—in this case, the HKHK region. Temperature varies primarily as a function of elevation. The novel temperature downscaling method used here derives monthly lapse rates empirically from low-spatial-resolution daily temperature data for five subregions within the HKHK, interpolates these monthly lapse rates to daily lapse rates, and then applies them at the spatial resolution of the elevation model used here (that is, approximately 900 meters). This process was conducted separately for the ERA-Interim and GCM simulation output for the historic period and 2040s, respectively.

Precipitation also varies as a function of elevation, but the relationship is based on the fact that clouds tend to precipitate as they move uphill, leaving less moisture in the atmosphere at higher elevations. Like temperature, this relationship is derived from the input data at a low resolution and then applied to the finer-scale elevation model. More details on the downscaling methodology selected for this book and a comparison of the selected method with other possible methods are presented in appendix C.

CCHF Model: Linking Climate, Snow and Glaciers, and Water Resources

The conceptual cryosphere hydrology framework assesses the impact of climate and BC emissions on the total availability of water and the distribution of water resources in the HKHK region in the 2040s relative to the reference historic period (1985–2015). CCHF is a mountain hydrology modeling framework that connects the inputs (climate, terrain) and physical hydrologic processes (snow accumulation, melt, runoff) to the resulting simulated hydrology (snowpack, glacier change, water availability). It is fully distributed over the spatial area set by the user and at the resolution of the inputs.

The model was calibrated with streamflow observations, geodetic mass balance estimates, and snow cover satellite imagery at a spatial resolution of 1 kilometer and daily time step (see appendix D). The model was validated with additional observations from these sites and subsequently used to produce the historic period (1985–2015) for the entire HKHK region. The results for the last full decade constitute the baseline for analyzing the impacts of climate and BC emissions. Figure 4.2 illustrates the model's structure.

CCHF has a modular structure with representations for 22 physical processes related to the cryosphere (shown on the left) and hydrology (shown on the right) in figure 4.2. The key processes used here and their representation are summarized in table 4.3. For example, the simulations use an "enhanced temperature index" representation of snow and glacier melt in which solar radiation, snow and ice albedo, and atmospheric transmittance are incorporated into the overall mountain hydrology representation.[5]

The modules of the CCHF used here include all major cryospheric and hydrologic processes. The process representations for these modules are also limited to those that depend directly on temperature, precipitation, and BC input data. The specific model was set up using a daily time step at a 900 meters x 900 meters spatial resolution. The modules and parameters were selected based on the skill of the respective models in replicating historic time-series streamflow and ice mass balance measurements and remote-sensed snow-covered area data at eight sub-basins in the HKHK region. The calibrated parameters were validated at these same sites over time periods not covered in the calibration period. All CCHF simulations were run for "water years," which begin on October 1, run through September 30, and are named based on the ending year.

Future water production projections were based on CCHF simulations that use the 2040s climate and BC described previously. The CCHF simulations used here were initialized for October 1, 1986, and simulated historic conditions with downscaled

FIGURE 4.2 Modular Structure of the Conceptual Cryosphere Hydrology Framework

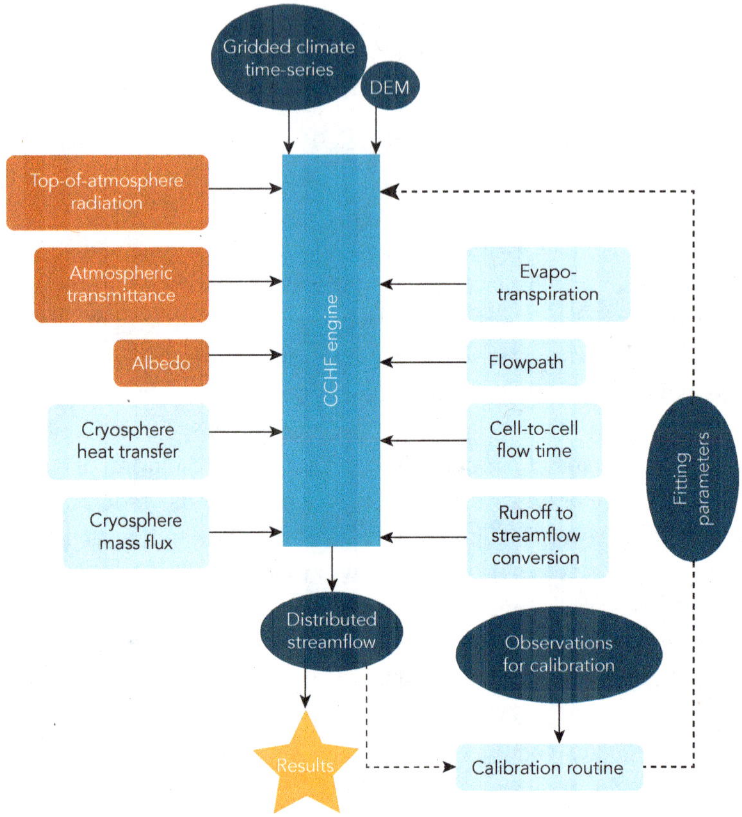

Source: Mosier, Hill, and Sharp 2016.
Note: This figure presents categories of process modules included in version 1 of the conceptual cryosphere hydrology framework (CCHF). Rounded blocks refer to process modules, and ovals refer to data inputs and outputs. Blocks tinted light blue denote modules used for all models, and blocks tinted orange denote modules only used for models that include shortwave radiation. CCHF = conceptual cryosphere hydrology framework. DEM = digital elevation model.

ERA-Interim climate and historic representative BC data through September 30, 2015. From October 1, 2015, through September 30, 2039, a "bridge run" of climate data based on the Community Climate System Model (CCSM) was used to connect the historic period to the 2040s of interest from a research perspective. The two 2040s ensembles composed of eight GCMs representing RCP 4.5 (standard) and RCP 4.5 (mitigation), with respective representative BC deposition amounts, were then run for the period October 1, 2039, through September 30, 2040. Comparing the results of these two 2040s ensembles and the historic baseline makes it possible to answer this book's fundamental research question: Would regional aerosol and BC policy affect glaciers and mountain water resources in the context of global climate change caused by GHGs?

TABLE 4.3 Process Representations Used in the Conceptual Cryosphere Hydrologic Framework (CCHF)

Process	Representation	Short explanation
Heat transfer	Enhanced temperature index (ETI) with empirical glacier debris	Main ETI formulation follows Pellicciotti et al. (2005). Impact of debris cover on ice heat transfer is an empirical representation derived from Kraaijenbrink et al. (2017).
Snow mass flux	Step based on temperature	Melt is allowed to occur when temperature is above an empirically derived value.
Ice flux	Ratio	Heat first melts any snowpack present. If there is additional heat, the portion of the heat that melts ice is based on how much has been used to melt snow. This scales the modeled ice heat flux.
Snow albedo	Amount of incoming shortwave radiation that is absorbed at the snow or ice surface	Exerts a strong influence on the spatial and temporal evolution of melt rates. Based on Pellicciotti et al. (2005).
Ice albedo	Constant	Debris-free and debris-covered albedos are based on broadly accepted values (DeWalle and Rango 2008).
Black carbon impact on albedo	Concentrations in snow and ice	Empirically derived formula that modifies snow and ice albedo based on field experiments by Ming et al. (2009).
Snowpack moisture-holding capacity	Percentage	The snowpack is modeled to hold an amount of water as an empirically derived portion of its total water-equivalent snow content.
Snow sublimation	Elevation dependent	Follows formulation by Lutz et al. (2016).
Snow to ice	Depth	Snow is assumed to become part of the ice when its depth is greater than a threshold.
Snow avalanche	Angle	Snow avalanches for all spatial grid cells when the steepest angle is greater than a threshold.
Runoff	Leaky bucket	The top layer of ground acts as a "leaky bucket" with a maximum reservoir size and a set percentage that drains every time step. Based on Moore, Trubilowicz, and Buttle (2012).
Top of atmosphere radiation	DeWalle and Rango method	Solar radiation is modeled as a function of day of the year and grid attributes (latitude, aspect, slope). Based on DeWalle and Rango (2008).
Atmospheric transmittance	Linear function of elevation	More top of atmosphere radiation reaches the ground the higher the elevation. Based on Coops, Wulder, and Iwanicka (2009).

Source: The explanations contain the sources for each process.

Notes

1. The computational time to evaluate the gridded model is inversely proportional to the square of the horizontal grid resolution.
2. Glacier debris plays an important and varying role in the energy balance of glaciers in South Asia (Mihalcea et al. 2008; Scherler, Bookhagen, and Strecker 2011). In very thin layers, debris increases melt by lowering the ice albedo (Lejeune et al. 2013; Reid and Brock 2010). In thick layers, glacier debris acts as an insulator and tends to reduce glacier melt.
3. Mani et al. (2018) review 18 global climate models of the Coupled Model Intercomparison Project (CMIP5) and select 11 based on their performance in modeling historic South Asian climate. They then use these 11 climate models as an ensemble to project long-term changes in average temperature and precipitation in the region.
4. For the RCP database, see http://tntcat.iiasa.ac.at/RcpDb/dsd?Action=htmlpage&page=about.
5. Alternative representations for mountain hydrology in CCHF include a "simple degree-index" representation of snow and glacier melt in which melt is a linear function of mean daily temperature.

References

Alvarado, M. J., E. Winijkul, R. Adams-Selin, E. Hunt, C. Brodowski, C. R. Lonsdale, D. T. Shindell, et al. 2018. "Sources of Black Carbon Deposition to the Himalayan Glaciers in Current and Future Climates." *Journal of Geophysical Research: Atmospheres* 123 (14): 7482–505. doi:10.1029/2018JD029049.

Arora, V. K., J. F. Scinocca, G. J. Boer, J. R. Christian, K. L. Denman, G. M. Flato, V. V. Kharin, W. G. Lee, and W. J. Merryfield. 2011. "Carbon Emission Limits Required to Satisfy Future Representative Concentration Pathways of Greenhouse Gases." *Geophysical Research Letters* 38 (5): L05805. doi:10.1029/2010GL046270.

Coops, N. C., M. A. Wulder, and D. Iwanicka. 2009. "Large Area Monitoring with a MODIS-Based Disturbance Index (DI) Sensitive to Annual Seasonal Variations." *Remote Sensing of Environment* 113 (6): 1250–61.

Dee, D. P., S. M. Uppala, A. J. Simmons, P. Berrisford, P. Poli, S. Kobayashi, U. Andrae, M. A. Balmaseda, G. Balsamo, and P. Bauer. 2011. "The ERA-Interim Reanalysis: Configuration and Performance of the Data Assimilation System." *Quarterly Journal of the Royal Meteorological Society* 137 (656): 553–97.

DeWalle, D., and A. Rango. 2008. *Principles of Snow Hydrology*. Cambridge, UK: Cambridge University Press.

Dufresne, J.-L., M. A. Foujols, S. Denvil, A. Caubel, O. Marti, O. Aumont, Y. Balkanski, et al. 2013. "Climate Change Projections Using the IPSLCM5 Earth System Model: From CMIP3 to CMIP5." *Climate Dynamics* 40 (9–10): 2123–65. doi:10.1007/s00382-012-1636-1.

Ekman, A. M. L. 2013. "Do Sophisticated Parameterizations of Aerosol-Cloud Interactions in CMIP5 Models Improve the Representation of Recent Observed Temperature Trends?" *Journal of Geophysical Research: Atmospheres* 119 (2): 817–32. doi:10.1002/2013JD020511.

Farr, T. G., P. Rosen, E. Caro, R. Crippen, R. Duren, S. Hensley, M. Kobrick, et al. 2007. The Shuttle Radar Topography Mission." *Reviews of Geophysics* 45 (2). doi:10.1029/2005RG000183.

Gent, P. R., G. Danabasoglu, L. J. Donner, M. M. Holland, E. C. Hunke, S. R. Jayne, D. M. Lawrence, et al. 2011. "The Community Climate System Model Version 4." *Journal of Climate* 24 (19): 4973–91.

GFDC Global Atmospheric Model Development Team. 2004. "The New GFDL Global Atmosphere and Land Model AM2–LM2: Evaluation with Prescribed SST Simulations." *Journal of Climate* 17 (24): 4641–73.

Giorgetta, M. A., J. Jungclaus, C. H. Reick, S. Legutke, J. Bader, M. Böttinger, V. Brovkin, et al. 2013. "Climate and Carbon Cycle Changes from 1850 to 2100 in MPI-ESM Simulations for the Coupled Model Intercomparison Project Phase 5: Climate Changes in MPI-ESM." *Journal of Advances in Modeling Earth Systems* 5 (3): 572–97. doi:10.1002/jame.20038.

Grell, G. A., S. E. Peckham, R. Schmitz, S. A. McKeen, G. Frost, W. C. Skamarock, and B. Elder. 2005. "Fully Coupled 'Online' Chemistry within the WRF Model." *Atmospheric Environment* 39 (37): 6957–75.

Hourdin, F., T. Mauritsen, A. Gettelman, J.-C. Golaz, V. Balaji, Q. Duan, D. Folini, et al. 2017. "The Art and Science of Climate Model Tuning." *Bulletin of the American Meteorological Society* 98 (3): 589–602. doi:10.1175/BAMS-D-15-00135.1.

IPCC (Intergovernmental Panel on Climate Change). 2013a. *Climate Change 2013: The Physical Science Basis; Working Group I Contribution to the Fifth Assessment Report of the Intergovernmental Panel on Climate Change,* edited by T. F. Stocker, D. Qin, G.-K. Plattner, M. Tignor, S. K. Allen, J. Boschung, A. Nauels, Y. Xia, V. Bex, and P. M. Midgley. New York: Cambridge University Press. doi:10.1017/CBO9781107415324.

IPCC (Intergovernmental Panel on Climate Change). 2013b. "Summary for Policymakers." In *Climate Change 2013: The Physical Science Basis; The Working Group I Contribution* to the *Fifth Assessment Report of the Intergovernmental Panel on Climate Change,* edited by T. F. Stocker, D. Qin, G.-K. Plattner, M. Tignor, S. K. Allen, J. Boschung, A. Nauels, Y. Xia, V. Bex, and P. M. Midgley. New York: Cambridge University Press. https://www.ipcc.ch/report/ar5/wg1/.

Jarvis, A., E. Guevara, H. L. Reuter, and A. D. Nelson. 2008. "Hole-Filled SRTM for the Globe: Version 4: Data Grid." CGIAR Consortium for Spatial Information, Montpellier. http://srtm.csi.cgiar.org/.

Kirkevåg, A. T., T. Iversen, Ø. Seland, C. Hoose, J. E. Kristjánsson, H. Struthers, A. M. L. Ekman, et al. 2013. "Aerosol-Climate Interactions in the Norwegian Earth System Model —NorESM1-M." *Geoscientific Model Development* 6: 207–44. doi:10.5194/gmd-6-207-2013.

Kraaijenbrink, P. D. A., M. F. P. Bierkens, A. F. Lutz, and W. W. Immerzeel. 2017. "Impact of a Global Temperature Rise of 1.5 Degrees Celsius on Asia's Glaciers." *Nature* 549 (7671): 257–60. doi:10.1038/nature23878.

Lejeune, Y., J. Bertrand, P. Wagnon, and S. Morin. 2013. "A Physically Based Model of the Year-Round Surface Energy and Mass Balance of Debris-Covered Glaciers." *Journal of Glaciology* 59 (214): 327–44. doi:10.3189/2013JoG12J149.

Lohmann, U., L. Rotstayn, T. Storelvmo, A. Jones, S. Menon, J. Quaas, A. M. L. Ekman, D. Koch, and R. Ruedy. 2010. "Total Aerosol Effect: Radiative Forcing or Radiative Flux Perturbation?" *Atmospheric Chemistry and Physics* 10 (7): 3235–46. doi:10.5194/acp-10-3235-2010.

Lutz, A. F., W. W. Immerzeel, P. D. A. Kraaijenbrink, A. B. Shrestha, and M. F. P. Bierkens. 2016. "Climate Change Impacts on the Upper Indus Hydrology: Sources, Shifts, and Extremes." *PLOS One* 11 (11): e0165630. doi:10.1371/journal.pone.0165630.

Mani, M., S. Bandyopadhyay, S. Chonabayashi, A. Markandya, and T. Mosier. 2018. *South Asia's Hotspots: The Impact of Temperature and Precipitation Changes on Living Standards.* South Asia Development Matters. Washington, DC: World Bank.

Mihalcea, C., C. Mayer, G. Diolaiuti, C. D'Agata, C. Smiraglia, A. Lambrecht, E. Vuillermoz, and G. Tartari. 2008. "Spatial Distribution of Debris Thickness and Melting from Remote-Sensing and Meteorological Data, at Debris-Covered Baltoro Glacier, Karakoram, Pakistan." *Annals of Glaciology* 48 (1): 49–57. doi:10.3189/172756408784700680.

Ming, J., C. Xiao, H. Cachier, D. Qin, X. Qin, Z. Li, and J. Pu. 2009. "Black Carbon (BC) in the Snow of Glaciers in West China and Its Potential Effects on Albedos." *Atmospheric Research* 92 (1): 114–23. doi:10.1016/j.atmosres.2008.09.007.

Moore, D., J. W. Trubilowicz, and J. M. Buttle. 2012. "Prediction of Streamflow Regime and Annual Runoff for Ungauged Basins Using a Distributed Monthly Water Balance Model." *JAWRA (Journal of the American Water Resources Association)* 48 (1): 32–42. doi:10.1111/j.1752-1688.2011.00595.x.

Mosier, T., D. F. Hill, and K. V. Sharp. 2016. "How Much Cryosphere Model Complexity Is Just Right? Exploration Using the Conceptual Cryosphere Hydrology Framework." *The Cryosphere* 10 (5): 2147. doi:10.5194/tc-10-2147-2016.

Mosier, T., D. F. Hill, and K. V. Sharp. 2017. "Update to the Global Climate Data Package: Analysis of Empirical Bias Correction Methods in the Context of Producing Very High-Resolution Climate Projections." *International Journal of Climatology* 38 (2): 825–40. doi:10.1002/joc.5213.

Pellicciotti, F., B. Brock, U. Strasser, P. Burlando, M. Funk, and J. Corripio. 2005. "An Enhanced Temperature-Index Glacier Melt Model Including the Shortwave Radiation Balance: Development and Testing for Haut Glacier d'Arolla, Switzerland." *Journal of Glaciology* 51 (175): 573–87.

Reid, T., and B. Brock. 2010. "An Energy-Balance Model for Debris-Covered Glaciers Including Heat Conduction through the Debris Layer." *Journal of Glaciology* 56 (199): 903–16.

RGI Consortium. 2017. "Randolph Glacier Inventory (RGI)—A Dataset of Global Glacier Outlines: Version 6.0; Technical Report." Global Land Ice Measurements from Space, Boulder, CO. doi:10.7265/N5-RGI-60.

Scherler, D., B. Bookhagen, and M. Strecker. 2011 "Spatially Variable Response of Himalayan Glaciers to Climate Change Affected by Debris Cover." *Nature Geoscience* 4: 156–59. doi:10.1038/ngeo1068.

Schmidt, G. A., M. Kelley, L. Nazarenko, R. Ruedy, G. L. Russell, I. Aleinov, M. Bauer, et al. 2014. "Configuration and Assessment of the GISS Model E2 Contributions to the CMIP5 Archive." *Journal of Advances in Modeling Earth Systems* 6 (1): 141–84. doi:10.1002/2013MS000265.

Shindell, D., and G. Faluvegi. 2009. "Climate Response to Regional Radiative Forcing during the Twentieth Century." *Nature Geoscience* 2 (4): 294.

Shindell, D. T., A. Voulgarakis, G. Faluvegi, and G. Milly. 2012. "Precipitation Response to Regional Radiative Forcing." *Atmospheric Chemistry and Physics* 12 (15): 6969–82.

Sperber, K. R., H. Annamalai, I.-S. Kang, A. Kitoh, A. Moise, A. Turner, B. Wang, and T. Zhou. 2013. "The Asian Summer Monsoon: An Intercomparison of CMIP5 vs. CMIP3 Simulations of the Late 20th Century." *Climate Dynamics* 41 (9–10): 2711–44.

Stohl, A., B. Aamaas, M. Amann, L. H. Baker, N. Bellouin, T. K. Berntsen, O. Boucher, et al. 2015. "Evaluating the Climate and Air Quality Impacts of Short-Lived Pollutants." *Atmospheric Chemistry and Physics* 15 (18): 10529–66.

Stott, P. A. 2003. "Attribution of Regional-Scale Temperature Changes to Anthropogenic and Natural Causes." *Geophysical Research Letters* 30 (14).

UNEP (United Nations Environment Programme). 2011. *Towards a Green Economy: Pathways to Sustainable Development and Poverty Eradication; A Synthesis for Policymakers.* Nairobi: UNEP.

Voldoire, A., E. Sanchez-Gomez, D. Salas y Mélia, B. Decharme, C. Cassou, S. Sénési, S. Valcke, et al. 2012. "The CNRM-CM5.1 Global Climate Model: Description and Basic Evaluation." *Climate Dynamics* 40: 2091–121. doi:10.1007/s00382-011-1259-y.

Wang, C. 2013. "Impact of Anthropogenic Absorbing Aerosols on Clouds and Precipitation: A Review of Recent Progresses." *Atmospheric Research* 122 (March): 237–49. doi:10.1016/j.atmosres.2012.11.005.

Wang, C. 2015. "Anthropogenic Aerosols and the Distribution of Past Large-Scale Precipitation Change." *Geophysical Research Letters* 42 (24): 10876–84. doi:10.1002/2015GL066416.

Watanabe, S., T. Hajima, K. Sudo, T. Nagashima, T. Takemura, H. Okajima, T. Nozawa, et al. 2011. "MIROC-ESM 2010: Model Description and Basic Results of CMIP5-20c3m Experiments." *Geoscientific Model Development* 4 (4): 845–72. doi:10.5194/gmd-4-845-2011.

Winijkul, E., and T. Bond. 2016. "Emissions from Residential Combustion Considering End-Uses and Spatial Constraints: Part II, Emission Reduction Scenarios." *Atmospheric Environment* 124 (pt. A): 1–11. doi:10.1016/j.atmosenv.2015.10.011.

Winijkul, E., L. Fierce, and T. C. Bond. 2016. "Emissions from Residential Combustion Considering End-Uses and Spatial Constraints: Part I, Methods and Spatial Distribution." *Atmospheric Environment* 125 (pt. A): 126–39.

CHAPTER 5

Modeling the Role of Black Carbon in Glacier Melt

Understanding the sources and transport of air pollution in South Asia is important because black carbon (BC) is becoming a significant factor in the retreat of some Himalayan glaciers (Bond et al. 2013; Ramanathan and Carmichael 2008; World Bank and ICCI 2013). In addition to threats from global climate change, BC produced and circulated within the region is both increasing the absorption of solar radiation by glaciers, through decreasing snow and glacier surface reflectance after deposition (Flanner et al. 2008; Menon et al. 2010), and raising air temperatures above the snow, which also increases melt. For this reason, the transport of air pollution can play a significant role in determining melting and hydrologic flows in South Asia.

While the representative concentration pathways (RCPs) span a wide range of total radiative forcing, they do not cover the full range of emissions, particularly for aerosols (IPCC 2013). Unlike greenhouse gases (GHGs), BC is more heterogenous and depends on regional and transport conditions. It has generally not been modeled at a large scale over long periods of time. The research for this book was formulated jointly with a complementary study that focused specifically on understanding the current sources and deposition of BC in the Himalayan glaciers. That study developed two scenarios of BC deposition in the 2040s. The first scenario assumes that countries in the region take no further control measures beyond their current commitments. The second scenario assumes that countries implement all available measures with beneficial impacts on air quality and climate.

Evidence suggests that BC contributes more to warming at higher altitudes. Xu, Lamarque, and Sanderson (2018) use modeling simulations to confirm the role of BC in the observed altitude dependence of warming in the Himalayan region, as described in chapter 2.[1] They find that the impacts of BC (atmospheric warming and surface darkening of snow) are coupled with positive snow-albedo feedback in the Himalayan region to account for the disproportionately large role of BC in high-elevation regions. In fact, studies have shown that BC concentrations in high-elevation Asian glaciers depend primarily on elevation—that is, higher sites have lower BC concentrations—and secondarily on the intensity of regional emissions, precipitation, and snow melting conditions (Qian et al. 2015). For these reasons, somewhat counterintuitively, while BC concentration declines with altitude, the influence of BC on warming increases due to the amplification effects discussed in this chapter. Therefore, when predicting future melt rates in the various Himalaya, Karakoram, and Hindu Kush (HKHK) watersheds, it is critical to factor the transport and deposition of air pollution into the overall modeling framework appropriately.

BC deposition records retrieved from Tibetan glaciers have shown that the mean BC mixing ratio in snow has increased significantly, from 20 nanograms per gram (or parts per billion) throughout the second half of the 20th century to 50 nanograms per gram after 1990 (Ming et al. 2008). Trends in the BC mixing ratio are also attributed to variations in atmospheric transport by summer monsoons, suggesting—again—that regional air pollution and transport affect melting and hydrological flows in South Asia.

Black Carbon and Air Pollution

South Asia is at the global epicenter of the generation and adverse impacts of air pollution. Most of the population in the region lives in places where air quality fails to meet the World Health Organization's guideline limit—10 micrograms per cubic meter a year (Amann et al. 2016; Mani and Yamada 2021). A multiplicity of sources and geographic source regions, modes of exposure, and impacts all add to the complexity of the air pollution problem in South Asia. For example, a recent assessment of sources of pollution in the Indian capital Delhi suggests that the combination of fuelwood and biomass burning, fuel adulteration, vehicle emissions, and large-scale burning of crop residue result in average annual concentrations of fine particulate matter ($PM_{2.5}$) reaching 120 micrograms per cubic meter or higher. The Greenhouse Gas-Air Pollution Interactions and Synergies (GAINS) model estimations in figure 5.1 show that BC and organic carbon constitute a significant proportion—between about 30 percent and 50 percent, respectively—of total $PM_{2.5}$ mass (Amann et al. 2016).

Air pollutants are sourced both locally and regionally. The level of air pollution is susceptible to the surrounding weather conditions, including wind velocity and direction. Even in densely populated cities like Delhi, 60 percent of the $PM_{2.5}$ concentration in ambient air is transported into the city from outside sources: of this percentage, half comes from the surrounding states of Haryana and Uttar Pradesh, a quarter from more remote anthropogenic sources, and a quarter from natural sources (Amann et al. 2016).

FIGURE 5.1 Share of Black Carbon and Organic Carbon in Total $PM_{2.5}$ Mass in Delhi, India

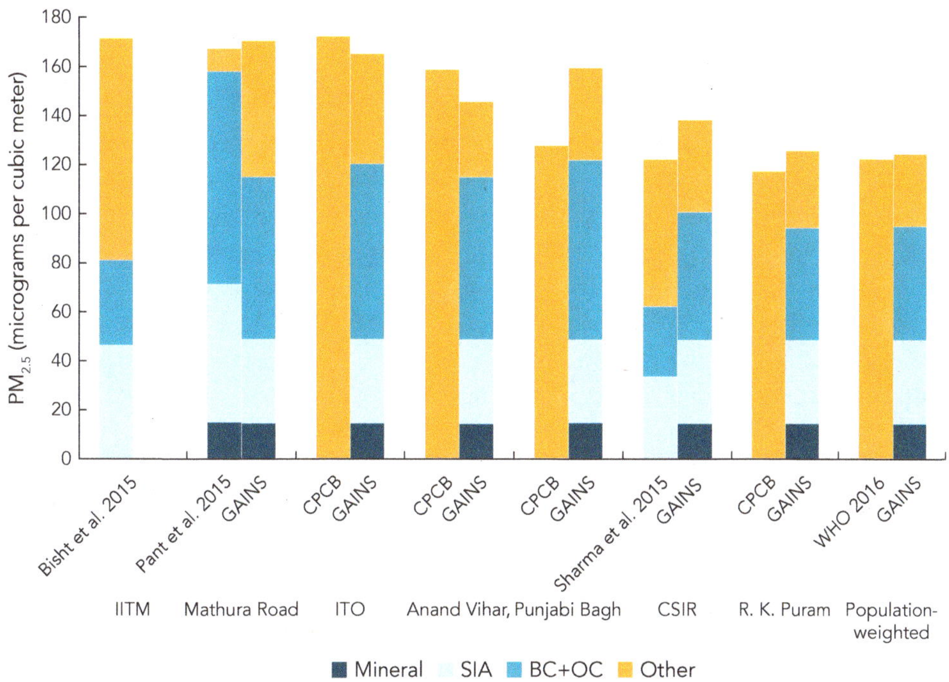

Source: Amann et al. 2016.

Note: The chart shows observations of $PM_{2.5}$ concentrations at the study sites compared to the chemical composition of $PM_{2.5}$, as estimated by the GAINS model results for 2015. The following are locations for monitoring stations: IITM (Indian Institute of Tropical Meteorology), Mathura Road, ITO (Income Tax Office), Anand Vihar, Punjabi Bagh, CPCB (Central Pollution Control Board), and R. K. Puram. BC = black carbon. CSIR = Council of Scientific & Industrial Research. GAINS = Greenhouse Gas-Air Pollution Interactions and Synergies. OC = organic carbon. $PM_{2.5}$ = fine particulate matter. SIA = secondary inorganic aerosol.

In areas with lower pollution-generating activities, such as areas within the HKHK region characterized by high elevation and hilly topography, a higher proportion of pollutants with BC and organic carbon is likely to be transported from external areas.

Evidence gathered to date suggests that the majority of BC transported to the HKHK comes from outside the region. While a lack of measurement data and significant uncertainty present in emissions inventories prevent a complete understanding of the sources of absorbing aerosols[2] over the snow and ice regions of the Himalayas, studies have developed an emerging picture of a region with strong gradients of deposition and impact and seasonal trends that are strongly influenced by monsoon dynamics (Gertler et al. 2016). Map 5.1 (map 5.1 is accessible in the Nontechnical Summary, appendix A, at https://openknowledge.worldbank.org/handle/10986/35600) shows the two BC emission inventories in 2005. These inventories clearly show very large source regions of BC emissions across the Gangetic plain and in eastern China. The climatic phenomena of western disturbance, which comes from Europe to the northern Indian states to cause rainfall and snow during winters, also carries BC aerosols from there. When the western disturbance reaches higher reaches of the Himalayas, the BC aerosols get deposited there, causing significant damage to the glaciers by artificially increasing the temperature.

In another study, Kopacz et al. (2011) use the GEOS-Chem[3] adjoint model to provide a spatially and seasonally adjusted estimate of the origins of BC arriving in the HKHK. The GEOS-Chem model provides a computationally robust estimate of the sensitivity of modeled concentrations to changes in emissions elsewhere in the modeling domain. In this way, the GEOS-Chem modeling approach provides a quantitative estimate of which source regions are contributing to observed BC over or deposited within the HKHK. Using this model, Kopacz et al. (2011) find that China and India contribute the overwhelming majority of BC transported to the region, with substantial seasonal contributions from the Middle East and Nepal. Finally, Kumar et al. (2015) use tagged tracers of BC within the WRF-Chem model to identify the dominant sources of BC in South Asia during the 2006 Integrated Campaign for Aerosols, Gases, and Radiation Budget, but they do not analyze the relative impacts on BC deposition.

Using long-term observations of BC over the Indian region, Manoj et al. (2019) show a decreasing trend in BC over the Himalayan region. However, it is not clear how representative the 13 sites used to establish the surface BC trend for all of India were. Also, it is not clear if those sites were located in rural background areas or dense urban environments, both of which are potentially influenced by local sources that may have gone up or down due to strictly local policies. The authors point out but do not definitively explain the apparent contradiction of a declining BC surface trend with the observed increase of BC throughout the total atmospheric column. Although the authors investigate whether the apparent contradiction might be due to increased vertical advection (uplift) over the period, they conclude that "the possible contribution to the free tropospheric aerosol absorption coming from uplifting of aerosols cannot be ruled out, but it is difficult to quantify without extensive modeling efforts." The study attempts to show some positive effects of policies undertaken on the ground to curb BC emissions, but it falls short of providing any conclusive evidence.

Black Carbon and Glacier Modeling to Date

Knowledge gaps persist despite studies on the influence of BC on glaciers. One important factor in modeling the hydrology in the Himalayas is sublimation, which results in snow and ice converting directly to water vapor rather than water. The distribution of sublimation is controlled primarily by wind speed; research on the Yala Glacier suggests that the fraction of snowfall returned to the atmosphere may be much higher in wind-exposed locations (Stigter et al. 2018).

Qian et al. (2015) find that, when collectively considering all light-absorbing particles (black carbon, brown carbon, and organic carbon over snow or ice), induced changes in snow albedo generate changes in surface-radiative flux of 5–25 watts per square meter during the spring, with a maximum in April or May. Other studies have shown smaller-magnitude forcings (1–3 watts per square meter by Ménégoz et al. 2014; 1.5 watts per square meter by Flanner et al. 2007). However, Ménégoz et al. (2014) acknowledge that the coarse global model employed is not able to resolve adequately the extreme variation in wet deposition of BC with altitude and includes periods without snow cover. Flanner et al. (2007) find much higher magnitudes (10–20 watts per square meter) when averaging only over the snow-covered region of the Himalayas, a forcing one to four times greater than that exerted by carbon dioxide (CO_2) alone.

Xu, Lamarque, and Sanderson (2018) use a global model with an approximately one-degree spatial resolution and scaling factors between two and four to bottom-up emissions inventories to reflect observations better. In addition to the altitude dependence of BC deposition described earlier, Xu et al. find that preindustrial to present-day increases in BC emissions reduce the annual average snow fraction over the Tibetan plateau by more than 6 percent (relatively) and reduce snow depth by approximately 19 percent. They also find that surface albedo decreases by more than 5 percent along the Himalayan mountain range and 1.4 percent over the entire Tibet region, providing positive local feedback to the enhanced local warming.

Kopacz et al. (2011), described above, calculate instantaneous radiative forcing (with and without the presence of BC particles in snow) of +3.78 to +15.6 watts per square meter at the five sites studied, with a minimum range in winter of approximately 3–11 watts per square meter across the sites and a maximum range in summer of approximately 7–16 watts per square meter. Their study does not account for the reduction in albedo due to dust or soil in the snow.

Results

A framework for assessing the impacts of black carbon on glaciers is necessarily complex. Given the literature reviewed in this chapter, it is clear that a robust framework for exploring the role of BC and other air pollution in influencing snow melt and the subsequent hydrology in the HKHK region must be able to account for the tremendous variation in elevation, the transport of pollution, and the surface dynamics of snow melt and reflectance. The direct radiative forcing of aerosols is critical over highly reflective snow surfaces like the Himalayas, where a small concentration of absorbing aerosols in the atmosphere can lead to significant warming. The magnitude of this direct radiative forcing strongly depends on the columnar aerosol loading (aerosol optical depth). This section outlines the methodology used in this book to develop the BC scenarios. The methodology and results are detailed in Alvarado et al. (2018).

As detailed in Alvarado et al. (2018), the BC study models the impact of BC in the HKHK under both current and future climate conditions. Specifically, it uses the Weather Research and Forecasting model coupled with the Chemistry model (WRF-Chem) to understand the impacts of regional emissions and transport on the wet and dry deposition of BC in the HKHK region at high spatial resolution under both current conditions and future climate projections for 2040–50 (Grell et al. 2005).[4] The BC study evaluates results using the WRF-Chem model for the same scenarios as the Goddard Institute for Space Studies (GISS) global climate model, which provides the driving boundary conditions for WRF-Chem and includes the RCP 4.5 (standard) and RCP 4.5 (mitigation) scenarios, but uses a 12-kilometer horizontal resolution grid that covers India and the HKHK region with 35 vertical layers (map 5.2) (map 5.2 is accessible in the Nontechnical Summary, appendix A, at https://openknowledge.worldbank.org/handle/10986/35600). The scenarios also include a sensitivity analysis of the three distinct phases of the El Niño–Southern Oscillation (ENSO) cycle.

The BC study employs tagged BC tracers from several source sectors and countries to assess their relative impact on BC deposition.[5] Tagged tracers for BC were added to the WRF-Chem model code following the methods used by Kumar et al. (2015); these tracers track BC aerosol emissions from five sectors (diesel fuel, industry, solid fuel, open burning, and biomass burning) and six nations (Bangladesh, China, India, Myanmar, Nepal, and Pakistan), as well as BC from the initial and boundary conditions. Then 28-day (14-day spin-up, 14-day analysis) WRF-Chem simulations were run for each season (January, winter monsoon; April, monsoon transition; July, summer monsoon; and October, monsoon transition) for both current conditions (represented using the moderate ENSO year of 2013) and three future years between 2040 and 2050—a moderate ENSO year, a La Niña year, and an El Niño year. A more detailed treatment of this methodology is provided in appendix B.

Results estimate total BC deposition in South Asia as the sum of the fluxes of wet and dry deposition of BC. Map 5.3 shows total BC deposition for the 14-day analysis period of each month simulated for 2013. As expected, these values are generally highest near high-emission regions and lower over the HKHK region. Specifically, BC deposition is higher in the western Karakoram and Hindu Kush ranges near the Pakistan-China and India-China borders, respectively, and lower over the Himalayas in the east. The seasonal cycle of deposition peaks in January over India, the Arabian Sea, and the Bay of Bengal, but the seasonal peak in deposition over the HKHK region varies, with the southern regions peaking from January to April and the northern regions peaking from July to October.

BC is sourced extensively from outside the HKHK. Map 5.4 shows the division of this BC deposition between anthropogenic sources within the modeling domain, biomass burning (wildfires) within the modeling domain, and the boundary conditions (and thus BC from all sources outside the domain). Since the analysis period begins only after a two-week spin-up has occurred, the fraction of deposition coming from the initial conditions is negligible everywhere.

MAP 5.3 Total Deposition of Black Carbon between the 15th and 29th of (a) January, (b) April, (c) July, and (d) October in South Asia, 2013

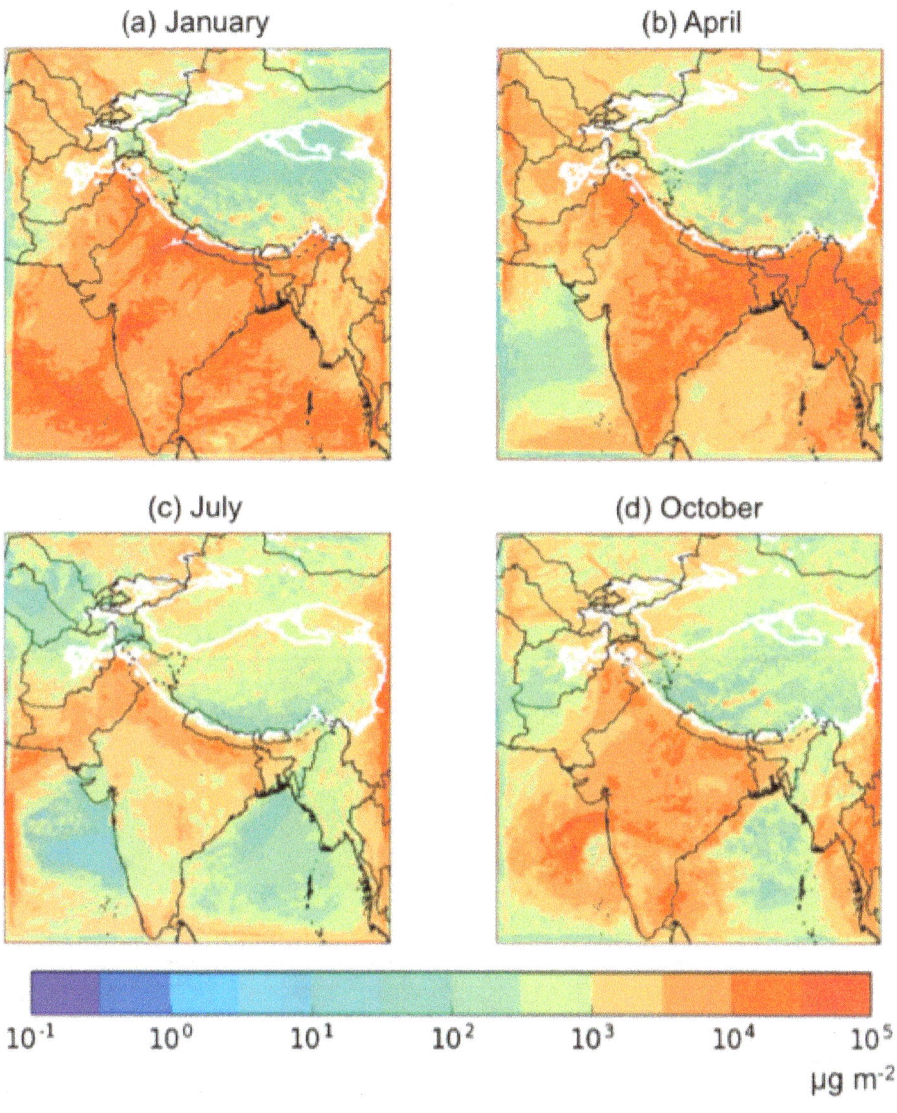

Source: Original map created for this publication, based on publicly available climate data sets.
Note: The thick white contour shows the boundary of the 700 hectopascal (hPa, millibar) surface pressure region. μg m^{-2} = micrograms per square meter.

As shown in map 5.4, anthropogenic sources of BC are responsible for most BC deposition over and near India regardless of season, while biomass burning–related BC peaks over Myanmar and Southeast Asia in April. BC from sources outside the domain specified in the modeling approach are consistently important in the northern

MAP 5.4 Sources of Black Carbon Deposition in the Himalaya, Karakoram, and Hindu Kush Region, by Month, 2013

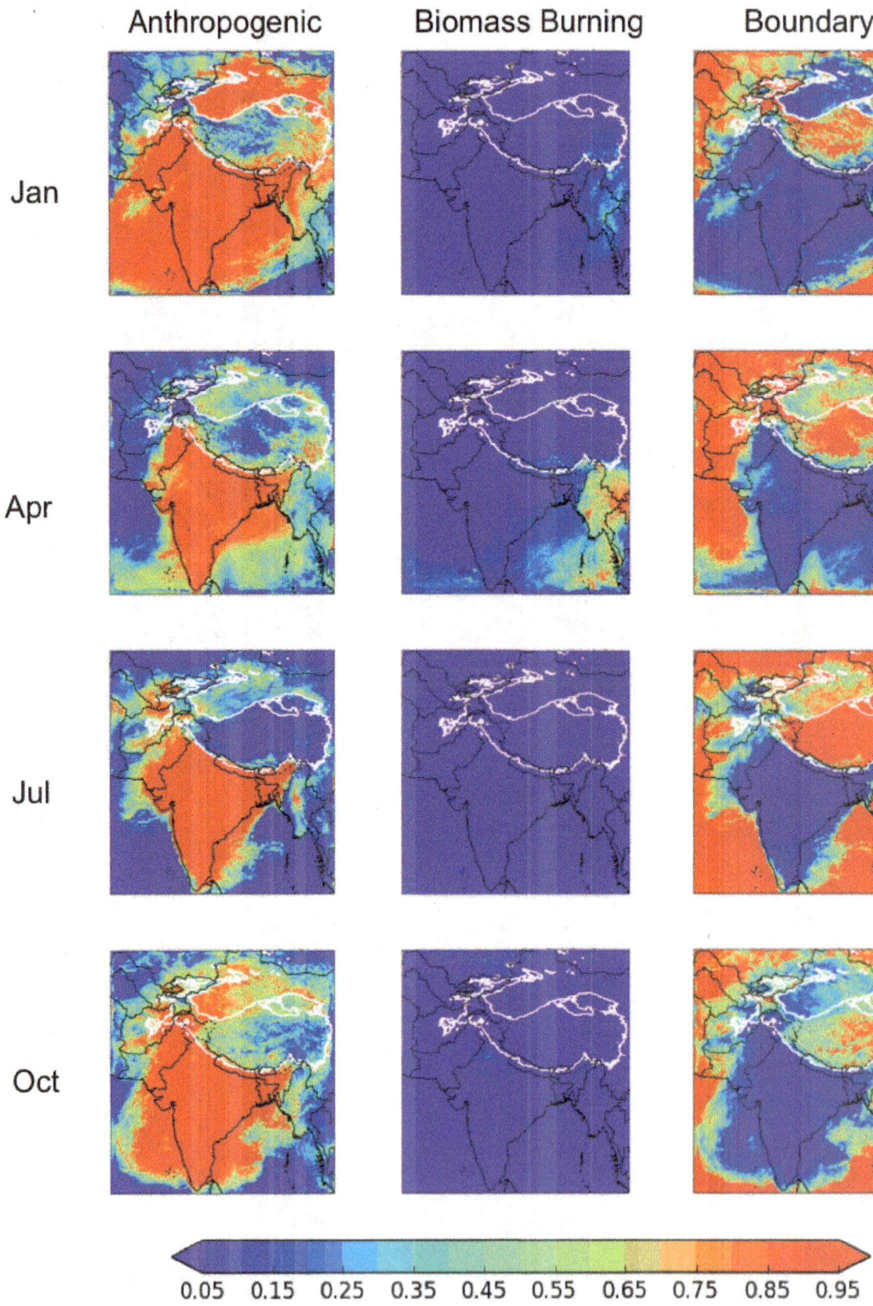

Source: Original map created for this publication, based on publicly available climate data sets.

part of the region, including strong impacts in the HKHK regions that peak during the summer monsoon in July. Results suggest that sources of BC outside of South Asia are responsible for a significant fraction of BC deposition in these regions, likely from the long-range transport of BC in the free troposphere that deposits to the mountains of the HKHK region. These results are also consistent with the global modeling results of Kopacz et al. (2011), who find that biomass burning in Africa and fossil fuel combustion in the Middle East can contribute significantly to the BC reaching the Himalayas and the Tibetan plateau.

Industrial and fuel processes are significant sources of BC in the region. Map 5.5 shows the fraction of in-domain anthropogenic (not total) BC deposition from each of the four major anthropogenic sources included in the BC study: industry (including brick kilns), diesel fuel, residential solid fuels, and open burning. These results may not be meaningful near the boundaries of the simulation, where boundary condition BC dominates the quantity of deposition. In the HKHK region, as shown, industry and solid fuel are important in-domain anthropogenic sources, with industry contributing between 31 and 41 percent (minimum of 8 milligrams per day in July, maximum of 46.4 milligrams per day in April) and solid fuel contributing 14–28 percent (minimum of 3.5 milligrams per day in July, maximum of 31.4 milligrams per day in January). Diesel fuel accounts for 7–10 percent of the HKHK BC deposition in January (8.1 milligrams per day), April (10.0 milligrams per day), and October (8.0 milligrams per day), but accounts for 18 percent in July (4.7 milligrams per day).

The BC study also estimates the fraction of in-domain anthropogenic deposition of BC that is due to anthropogenic sources in the six countries studied for each season. As expected, each country has a strong influence on anthropogenic BC deposition within its own jurisdiction, but there is also significant cross-boundary transport of BC in South Asia. For the HKHK region, Bangladesh and Myanmar generally have a negligible impact on cross-boundary BC, while China, India, Nepal, and Pakistan have noticeable contributions. The relative importance of the different nations also varies by season, with China and Pakistan contributing more in-domain anthropogenic deposition in July, while India's contribution peaks in October and January. However, as in-domain anthropogenic sources account for only 13–62 percent of the total BC deposition in this region, the relative contribution of each nation is generally well below 50 percent.

Overall, under a moderate ENSO year, sources outside the South Asian modeling domain have a similar impact on total BC deposition in the HKHK region (35–87 percent, varying by month) as South Asian anthropogenic sources (13–62 percent), with the boundary contribution peaking in July. The in-domain anthropogenic contribution is from industry (primarily brick kilns) and residential burning of solid fuel, which, when combined, account for 45–66 percent of the in-domain anthropogenic BC deposition in the HKHK region, with on-road diesel fuels making a smaller contribution (7–18 percent, peaking in July) and open burning accounting for less than 3 percent in all seasons. Other sources of BC from anthropogenic combustion (waste burning) were not explicitly tracked but account for the remaining fraction of in-domain emissions.

MAP 5.5 In-Domain Contributions to Black Carbon Deposition in the Himalaya, Karakoram, and Hindu Kush Region, 2013

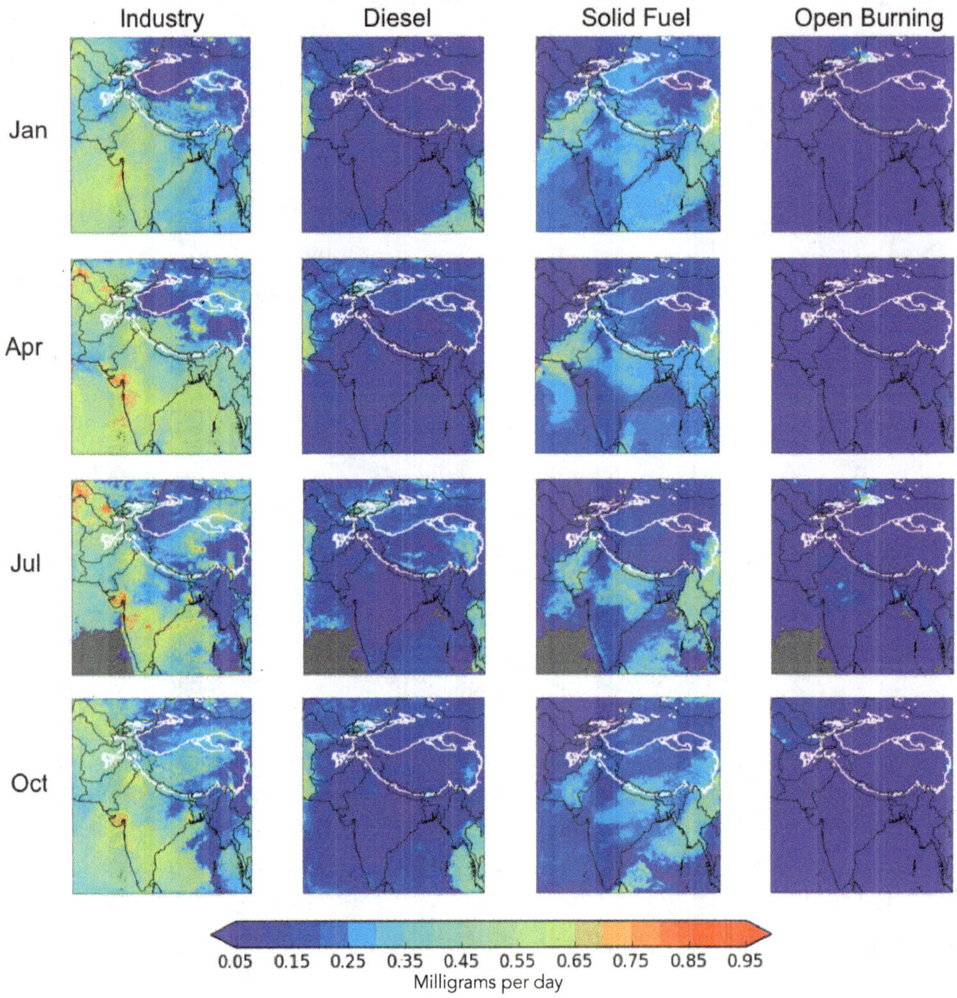

Source: Original map created for this publication, based on publicly available climate data sets.

Note: Fraction of total in-domain anthropogenic deposition of black carbon from (left to right) industry (including brick kilns), diesel fuel, solid fuel, and open (agricultural) burning in 2013. The thick white contour shows the boundary of the 700 hectopascal (hPa, 1 millibar) surface pressure region.

Notes

1. Liu et al. (2009) find a strong altitude dependence of surface warming, with peak warming trends of 2°C–2.5°C at 5,000 meters from 1961 to 2006.
2. While organic carbon is usually considered a reflective aerosol relative to typical underlying ground cover, the gray-white material is more energy and light absorbing than pristine white snow and thus will absorb energy along with BC in regions with snow and ice.
3. GEOS-Chem is a global 3-D model of atmospheric chemistry driven by meteorological input from the Goddard Earth Observing System (GEOS) of the NASA Global Modeling and Assimilation Office.
4. WRF-Chem is a coupled meteorological and chemical transport model driven by a physically explicit global climate and aerosol model with detailed emission inventories.
5. In this approach, BC from different source sectors is "tagged" by having the model simulate a set of species that have the physical and chemical properties of BC, but are emitted, transported, and deposited separately so that the impacts of the different sectors on the final BC deposition can be calculated.

References

Alvarado, M. J., E. Winijkul, R. Adams-Selin, E. Hunt, C. Brodowski, C. R. Lonsdale, D. T. Shindell, et al. 2018. "Sources of Black Carbon Deposition to the Himalayan Glaciers in Current and Future Climates." *Journal of Geophysical Research: Atmospheres* 123 (14): 7482–505. doi:10.1029/2018JD029049.

Amann, M., P. Purohit, I. Bertok, J. Borken-Kleefeld, J. Cofala, C. Heyes, G. Kiesewetter, et al. 2016. "Managing Future Air Quality in Delhi." International Institute for Applied Systems Analysis (IIASA), Laxenburg, Austria; National Environmental Engineering Research Institute (NEERI), Nagpur/Delhi, India. https://www.airqualityasia.org/pdf/2016-10-Managing-future-air-quality-in-Delhi.pdf.

Bisht, D. S., U. C. Dumka, D. G. Kaskaoutis, A. S. Pipal, A. K. Srivastava, V. K. Soni, S. D. Attiri, M. Sateesh, and S. Tiwari. 2015. "Carbonaceous Aerosols and Pollutants over Delhi Urban Environment: Temporal Evolution, Source Apportionment, and Radiative Forcing." *Science of the Total Environment* 521-522 (July 15): 431–45. doi:10.1016/j.scitotenv.2015.03.083.

Bond, T. C., S. J. Doherty, D. W. Fahey, P. M. Forster, T. Berntsen, B. J. DeAngelo, and C. S. Zender. 2013. "Bounding the Role of Black Carbon in the Climate System: A Scientific Assessment: Black Carbon in the Climate System." *Journal of Geophysical Research: Atmospheres* 118 (11): 5380–552. doi:10.1002/jgrd.50171.

Flanner, M. G., C. S. Zender, P. G. Hess, N. M. Mahowald, T. H. Painter, V. Ramanathan, and P. J. Rasch. 2008. "Spring-Time Warming and Reduced Snow Cover from Carbonaceous Particles." *Atmospheric Chemistry and Physics Discussion* 8: 19819–59.

Flanner, M. G., C. S. Zender, J. T. Randerson, and P. J. Rasch. 2007. "Present-Day Climate Forcing and Response from Black Carbon in Snow." *Journal of Geophysical Research: Atmospheres* 112 (D11202). doi:10.1029/2006JD008003, 2007.

Gertler, C. G., S. P. Puppala, A. Panday, D. Stumm, and J. Shea. 2016. "Black Carbon and the Himalayan Cryosphere: A Review." *Atmospheric Environment* 125 (pt. B): 404–17.

Grell, G. A., S. E. Peckham, R. Schmitz, S. A. McKeen, G. Frost, W. C. Skamarock, and B. Eder. 2005. "Fully Coupled "Online" Chemistry within the WRF Model." *Atmospheric Environment* 39 (37): 6957–975. doi:10.1016/j.atmosenv.2005.04.027.

IPCC (Intergovernmental Panel on Climate Change). 2013. "Summary for Policymakers." In *Climate Change 2013: The Physical Science Basis; The Working Group I Contribution* to the *Fifth Assessment Report of the Intergovernmental Panel on Climate Change,* edited by T. F. Stocker, D. Qin, G.-K. Plattner, M. Tignor, S. K. Allen, J. Boschung, A. Nauels, Y. Xia, V. Bex, and P. M. Midgley. New York: Cambridge University Press. https://www.ipcc.ch/report/ar5/wg1/.

Kopacz, M., D. L. Mauzerall, J. Wang, E. M. Leibensperger, D. K. Henze, and K. Singh. 2011. "Origin and Radiative Forcing of Black Carbon Transported to the Himalayas and Tibetan Plateau." *Atmospheric Chemistry and Physics* 11 (6): 2837–52. doi:10.5194/acp-11-2837-2011.

Kumar, R., M. Barth, V. S. Nair, G. G. Pfister, S. Suresh Babu, S. K. Satheesh, K. K. Moorthy, G. R. Carmichael, Z. Lu, and D. G. Streets. 2015. "Sources of Black Carbon Aerosols in South Asia and Surrounding Regions during the Integrated Campaign for Aerosols, Gases, and Radiation Budget (ICARB)." *Atmospheric Chemistry and Physics* 15 (10): 5415–28. doi:10.5194/acp-15-5415-2015.

Liu, X., Z. Cheng, L. Yan, and Z. Y. Yin. 2009. "Elevation Dependency of Recent and Future Minimum Surface Air Temperature Trends in the Tibetan Plateau and Its Surroundings: Global Planet." *Change* 68 (3): 164–74. doi:10.1016/j.gloplacha.2009.03.017.

Mani, M., and T. Yamada. 2021. "Exploring Air Pollution and COVID-19 Linkages in South Asia." Forthcoming in *Oxford Encyclopedia of Environmental Economics.*

Manoj, M. R., S. K. Satheesh, K. K. Moorthy, M. M. Gogoi, and S. S. Babu. 2019. "Decreasing Trend in Black Carbon Aerosols over the Indian Region." *Geophysical Research Letters* 46(5): 2903–10. doi:10.1029/2018GL081666.

Ménégoz, M., G. Krinner, Y. Balkanski, O. Boucher, A. Cozic, S. Lim, P. Ginot, et al. 2014. "Snow Cover Sensitivity to Black Carbon Deposition in the Himalayas." *Atmospheric Chemistry and Physics* 14: 4237–49.

Menon, S., D. Koch, G. Beig, S. Sahu, J. Fasullo, and D. Orlikowski. 2010. "Black Carbon Aerosols and the Third Polar Ice Cap." *Atmospheric Chemistry and Physics* 10: 4559–71.

Ming, J., H. Cachier, C. Xiao, D. Qin, S. Kang, S. Hou, and J. Xu. 2008. "Black Carbon Record Based on a Shallow Himalayan Ice Core and Its Climatic Implications." *Atmospheric Chemistry and Physics* 8 (5): 1343–52. doi:10.5194/acp-8-1343-2008.

Pant, P., A. Shukla, S. D. Kohl, J. C. Chow, J. G. Watson, and R. M. Harrison. 2015. "Characterization of Ambient PM2.5 at a Pollution Hotspot in New Delhi, India and Inference of Sources." *Atmospheric Environment* 190 (May): 178–89. doi:10.1016/j.atmosenv.2015.02.074.

Qian, Y., T. J. Yasunari, S. J. Doherty, M. G. Flanner, W. K. M. Lau, J. Ming, H. Wang, M. Wang, S. G. Warren, and R. Zhang. 2015. "Light-Absorbing Particles in Snow and Ice: Measurement and Modeling of Climatic and Hydrological Impact." *Advances in Atmospheric Sciences* 32 (1): 64–91. doi:10.1007/s00376-014-0010-0.

Ramanathan, V., and G. Carmichael. 2008. "Global and Regional Climate Changes due to Black Carbon." *Nature Geoscience* 1 (4): 221–27.

Sharma, S., A. Goel, D. Gupta, A. Kumal, A. Mishra, S. Kundu, S. Chatani, and Z. Klimont. 2015. "Emission Inventory of Non-Methane Volatile Organic Compounds from Antropogenic Sources in India." *Atmospheric Environment* 102 (February): 209–19. doi:10.1016/j.atmosenv.2014.11.070.

Stigter, E. E., M. Litt, J. F. Steiner, P. N. J. Bonekamp, J. M. Shea, M. F. Bierkens, and W. W. Immerzeel. 2018. "The Importance of Snow Sublimation on a Himalayan Glacier." *Frontiers in Earth Science* 6: 108.

WHO (World Health Organization). 2016. "Ambient Air Pollution: A Global Assessment of Exposure and Burden of Disease." WHO, Geneva, Switzerland.

World Bank and ICCI (International Cryosphere Climate Initiative). 2013. *On Thin Ice: How Cutting Pollution Can Slow Warming and Save Lives.* A Joint Report of the World Bank and the ICCI, Washington, DC.

Xu, Y., J.-F. Lamarque, and B. M. Sanderson. 2018. "The Importance of Aerosol Scenarios in Projections of Future Heat Extremes." *Climatic Change* 146 (3–4): 393–406. doi:10.1007/s10584-015-1565-1.

CHAPTER 6

Impacts on Mountain Water Availability

This study used the conceptual cryosphere hydrology framework (CCHF) to assess the impact of climate change and black carbon (BC) emissions on water production in the mountain headwaters of the Indus, Ganges, and Brahmaputra river basins. The fundamental experiment was to assess the relative impact of a "standard" representative concentration pathway (RCP) 4.5 scenario versus an aerosol and BC "mitigation" RCP 4.5 scenario for the 2040s (defined as water years 2040–49). The difference between these scenarios is the quantity of aerosols in the atmosphere affecting temperature and precipitation and the amount of BC deposited on snow and glacier surfaces affecting albedo. The results discussed in this chapter are the product of a relative sensitivity analysis rather than a comprehensive assessment with definitive uncertainty bounds around the future state of water.

Within this context, the study examined changes in the share of water contributed by glacier melt, snow melt, and rainfall runoff. All of the projection simulations were created using a version of the CCHF model calibrated and validated for the Himalaya, Karakoram, and Hindu Kush (HKHK) region. Calibration and validation were done using a set of eight calibration sites for which glacier mass balance, snow cover, and streamflow measurements were available. Snow cover imagery for the entire region from 2005 to 2015 was used for additional validation, because locational measurements are sparse within the region. In these results, the upstream area is defined as the portion of the basin at and above 2,000 meters. Given the focus of this study, the discussion in the rest of this chapter centers on the upstream areas (above 2,000 meters) of the Indus, Ganges, and Brahmaputra basins. While the entire historic baseline period extends from 1986 to 2015, comparisons of the 2040s simulations were done with water years 2000–09, a recent historic decade.

The results presented here are averages of daily model simulations over the water year for the period of interest. Additional calibration and validation details are outlined in appendix D.

Current HKHK Water Production

The decadal average water runoff and the relative contribution of each source during the reference period for the entire HKHK region are shown in table 6.1, based on the validation simulation runs. As noted in other studies, the upper Indus basin receives a larger portion of its runoff from glacier melt (10 percent compared to 3–4 percent for the upper Ganges and upper Brahmaputra). Similarly, the contribution of snow melt to runoff is highest in the upper Indus. The Indus is higher elevation, colder, and drier than the other two upper basins. Despite this, it receives some rainfall (accounting for 35 percent of its total runoff), while the Ganges and Brahmaputra receive 66–70 percent of their total runoff from rain.

While there is considerable uncertainty regarding water production in the HKHK region, the results for total runoff in this study are consistently higher than those in prior studies. For example, estimates of the total production of runoff in two notable recent studies—Lutz et al. (2014) and Armstrong et al. (2019)—for these three basins differ by 15–90 percent of each other (table 6.2). Methodological differences such as the use of imagery and isotope analysis in Armstrong et al. (2019) and a simple degree-index model in Lutz et al. (2014) can explain some of the difference. However, precise comparisons of the results are confounded by differences in the spatial extent and reference periods used in each of the studies. All three studies use the same elevation threshold of 2,000 meters to distinguish between upstream and downstream portions of the basins.

The partitioning of runoff between ice melt, snow melt, and rainfall runoff in this study lies between prior estimates. There is greater uncertainty in the relative contributions of glacier melt, snow melt, and rain runoff (table 6.3). For example, Lutz et al. (2014) estimate that 41 percent of runoff in the upper Indus is derived from glacier melt, while Armstrong et al. (2019) estimate 3 percent. The differences between these two previous studies are similarly large for the other basins. Notably, Armstrong et al. (2019) portray a region where snow matters most and glacier contributions matter very little. In contrast, Lutz et al. (2014) identify larger contributions from glaciers and rainfall, likely underestimating the importance of snow within the region. With respect to the role of glaciers, the results in this study are closer to those of Armstrong et al. (2019) than to those of Lutz et al. (2014).

Consistent with the preceding studies, rainfall runoff is the dominant source of water production in the upper Ganges and upper Brahmaputra basins. While snow and glacier melt provide a combined total of about 30 percent of the water production in these basins, about two-thirds is derived from rainfall runoff. By contrast,

TABLE 6.1 Water Runoff and Partitioning between Sources in the Himalaya, Karakoram, and Hindu Kush Region, by Basin, during the Reference Period (2000–10)

Basin	Annual precipitation (millimeters per year)	Annual runoff (millimeters per year)	Contribution to total runoff (%)		
			Glacier melt	Snow melt	Rainfall runoff
Upper Ganges	1,350	1,296	4	31	65
Upper Brahmaputra	1,981	1,932	3	27	70
Upper Indus	789	796	10	55	35

Source: Created for this publication based on publicly available climate data sets.

TABLE 6.2 Comparison of Runoff in the Himalaya, Karakoram, and Hindu Kush Region across Studies during the Historic Period
runoff (millimeters per year)

Basin	This study, 2000–09	Lutz et al. (2014), 1998–2007	Armstrong et al. (2019), 2001–14
Upper Ganges	1,296	1,088	729
Upper Brahmaputra	1,932	691	1,319
Upper Indus	796	574	657

Sources: Publicly available climate data sets; Armstrong et al. 2019; Lutz et al. 2014.

TABLE 6.3 Estimated Runoff and Partitioning between Sources in the Himalaya, Karakoram, and Hindu Kush Region, by Study
runoff partition (%)

Basin	Glacier melt			Snow melt		
	Research for this book	Lutz et al. (2014)	Armstrong et al. (2019)	Research for this book	Lutz et al. (2014)	Armstrong et al. (2019)
Upper Ganges	4	12	1	31	9	47
Upper Brahmaputra	3	16	<1	27	9	73
Upper Indus	10	41	3	55	22	73

Sources: Publicly available climate data sets; Armstrong et al. 2019; Lutz et al. 2014.

rainfall runoff contributes only about a third of the water production in the upper Indus basin, consistent with the findings of Lutz et al. (2014) and Armstrong et al. (2019) (map 6.1, [map 6.1 is accessible in the Nontechnical Summary, appendix A, at https://openknowledge.worldbank.org/handle/10986/35600].).

Elevation

Elevation is a key factor contributing to hydrology. The elevation of an area affects the temperature and precipitation patterns, which in turn condition the hydrology of the basin. Most of the Ganges basin lies downstream, while most of the Brahmaputra basin lies upstream (table 6.4).

Sensitivity of Temperature and Precipitation to Greenhouse Gases and Black Carbon

Considering the significant uncertainty in future climate projections, two ensembles of six global climate models (GCMs) were used to generate the climate inputs that drive the 2040s CCHF model simulations. One ensemble exactly reflects RCP 4.5, and the other was modified to reflect simulations conducted as part of this research to assess the impact of BC (and aerosols overall) on temperature and precipitation. The Goddard Institute for Space Studies (GISS) model was used to assess the impact of aerosols on weather patterns because it contains physics representations of the interaction between particulates and atmospheric dynamics. The impact of a reduction in aerosols on temperature and precipitation was applied to the ensemble of six GCMs as a quantile mapping bias correction step prior to downscaling. The methodology for these adjustments is outlined in chapter 4 and discussed in detail in the appendixes, as summarized in table 4.1. These adjustments provided a seamless data set for the historic period to the projected 2040s scenarios that recognizes the large variations in temperature and precipitation within small areas arising from their elevation and topography and is consistent with the BC scenarios used.

Long-term average temperature is expected to rise across the HKHK region, while the long-term average annual precipitation is expected to rise only in some areas. Differences between the RCP 4.5 (standard) and the historic baseline across the HKHK

TABLE 6.4 Elevation Characteristics in the Himalaya, Karakoram, and Hindu Kush Region, by Basin

Basin	Median elevation (meters)	Drainage area (square kilometers)	Upstream area (square kilometers)	Upstream share (%)
Ganges	242	767,957	179,036	23
Brahmaputra	3,487	490,692	294,606	60
Indus	1,666	861,724	383,230	44

Source: Created for this publication based on publicly available climate data sets.

region are shown in map 6.2 (map 6.2 is accessible in the Nontechnical Summary, appendix A, at https://openknowledge.worldbank.org/handle/10986/35600). for the annual and monsoon periods. The largest temperature increases in the study area occur in the high mountain areas (60°–110° E and 25°–40° N). This finding is consistent with observations of elevation-dependent warming (Rangwala and Miller 2012). While the differences in precipitation levels vary spatially, they remain consistent between seasons. The largest increases occur in the Ganges delta area (72°–90° E and 25°–32° N), with maximum increases of around 4 millimeters per day during the monsoon. Precipitation is also projected to increase moderately in southern India (up to approximately 2 millimeters per day during the monsoon).

The long-term average temperature in the region is higher by approximately 0.4° C in the RCP 4.5 (mitigation) compared to the RCP 4.5 (standard) scenario. However, the differences in long-term average temperature vary considerably between these scenarios, with the differences being smaller than the bounds of the confidence intervals. While uncertain, the projected differences are spatially coherent between seasons. The projected long-term average annual precipitation is similar between the RCP 4.5 (standard) and RCP 4.5 (mitigation) scenarios, with few locations showing differences outside of the confidence intervals. However, these localized patches are not spatially coherent across the seasons. Likewise, the differences in the range of long-term average monthly temperature between the RCP 4.5 (standard) and RCP 4.5 (mitigation) scenarios are small and fall within the confidence intervals for most areas in the region. Bangladesh, eastern India, and Myanmar are exceptions, showing an increase of 0.3°C in the range of long-term average monthly temperature for the RCP 4.5 (mitigation) scenario compared to the RCP 4.5 (standard) scenario (map 6.3, [map 6.3 is accessible in the Nontechnical Summary, appendix A, at https://openknowledge.worldbank.org/handle/10986/35600].).

Black Carbon Deposition in the Region

The modeled BC deposition data used as an input to the CCHF model simulations were developed from simulations of the Weather Research and Forecasting coupled with Chemistry (WRF-Chem) model, as summarized in chapter 5 and detailed in Alvarado et al. (2018). The scenarios seek to capture the total wet and dry deposition of BC during the historic period and in the two future scenarios RCP 4.5 (standard) and RCP 4.5 (mitigation) that are consistent with aerosol projections in the respective cases. These depositions are used in the CCHF model to simulate the direct impacts of BC on snow and glacier surface albedo. As noted, there is more snow than glaciers in each of the upper basins, but BC gets covered by fresh snow and blends with the soil layer once all of the snow melts. In contrast, BC accumulates on the glacier surface over multiple seasons, with a longer-lasting and potentially more locally profound impact on melt.

The WRF-Chem model simulations were conducted for four seasons in three different years in the historic period and in the 2040s period. The results for the three future years in the 2040s were not significantly different from each other, so a single year was chosen as representative of the decade. The three sets of seasonal results were interpolated to daily values and downscaled to the spatial resolution of the rest of the CCHF model. The levels and seasonal fluctuations of the BC depositions used to drive the CCHF simulations are summarized in figure 6.1.

BC deposition during the historic period is lowest during the summer rainy months but gradually ramps up to nearly triple the minimum points in the summer. If fully implemented, already enacted BC emission polices are projected to reduce total annual BC deposits 23 percent across the region. Implementing additional policy measures that are currently technically and economically feasible could reduce the BC deposits an additional 50 percent. The seasonality of deposition matters for snow since the snowpack "refreshes" when new snow falls and when snow melts; the seasonality of deposition on glacier surfaces matters less.

Water Production and Partition in the Region

The relative importance of policies to reduce BC (and aerosol) emissions in the context of a changing climate can be teased out from comparisons across the different

FIGURE 6.1 Deposition of Black Carbon Used for All Global Climate Models, by Month

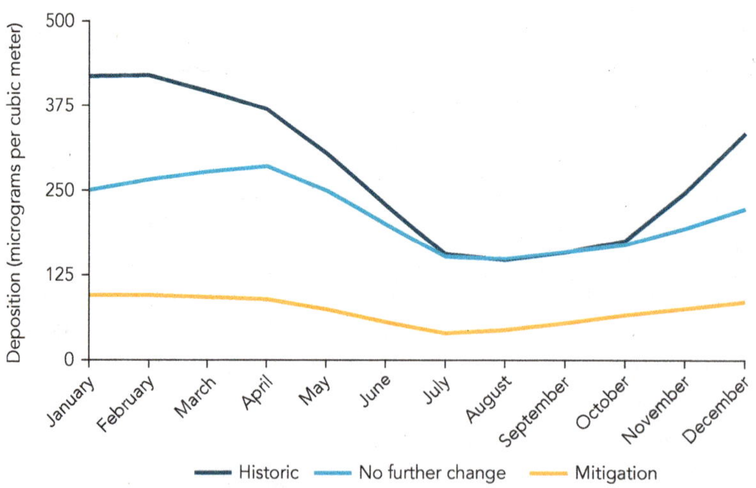

Source: Created for this publication based on publicly available climate data sets.

simulations, but the relative impacts of policies to reduce BC and greenhouse gas (GHG) emissions cannot be disentangled completely. The difference between the two RCP 4.5 scenarios is the quantity of aerosols in the atmosphere affecting temperature and precipitation as well as the amount of BC deposited on snow and glacier surfaces affecting albedo; the difference reflects the impacts of changes in BC policies in the region. Similarities in results between the two 2040s scenarios relative to the baseline are more attributable to changes in BC and GHG policies outside the region and globally. In addition, comparisons between the RCP 4.5 (standard) and the baseline provide a benchmark for assessing the impacts of climate change.

RCP 4.5 (STANDARD) COMPARED TO THE HISTORIC AVERAGE

The ensemble mean decadal average water runoff in the RCP 4.5 (standard) scenario is about 17 percent higher in the upper Indus compared to the baseline and is approximately the same in the other two basins. The decadal average water runoff and its partition between glacier melt, snow melt, and rainfall runoff for the entire HKHK region are shown in map 6.4, (map 6.4 is accessible in the Nontechnical Summary, appendix A, at https://openknowledge.worldbank.org/handle/10986/35600) and summarized for the three basins in table 6.5. Higher precipitation levels and the commensurate increases in rainfall partially account for the increased production of water.

Glacier melt contributes more to the overall water resources for all three basins in absolute amounts and as a share of the total water runoff in the RCP 4.5 scenario relative to baseline. Water released from glacier melt is double the levels in the reference case for the Indus and the Brahmaputra and triple the level in the Ganges. Glacier melt is projected to contribute a larger share of water production in the RCP 4.5 scenario compared to the baseline, substituting largely for snow melt in all three basins. Decreases in snow melt occur along the lower elevation areas of the upper Ganges and Indus river basins, suggesting that more precipitation is falling as rainfall rather than snow as higher temperatures cross the water vapor threshold. This

TABLE 6.5 Water Runoff and Partitioning between Sources in the Himalaya, Karakoram, and Hindu Kush Region, by Basin (RCP 4.5 Standard Scenario)

Basin	Annual precipitation (millimeters)	Annual runoff (cubic kilometers per year)	Contribution to total runoff (%)		
			Glacier melt	Snow melt	Rainfall runoff
Upper Ganges	1,246	1,329	11	20	69
Upper Brahmaputra	1,927	1,915	5	22	73
Upper Indus	846	932	16	45	39

Source: Created for this publication based on publicly available climate data sets.

situation arises simply because a few degrees of warming is sufficient to cross the water vapor threshold.

These changes are consistent across the six GCMs, with relatively small variations in the differences in total runoff and the contributions of glacier melt, snow melt, and rainfall runoff. The projected decadal means of total water runoff, rainfall runoff, and water released from snow melt and ice melt from the six GCMs are summarized in map 6.5 (map 6.5 is accessible in the Nontechnical Summary, appendix A, at https://openknowledge.worldbank.org/handle/10986/35600) and figure 6.2. Greater variations in projected precipitation across the GCMs percolate to larger variations in rainfall runoff. The models' projections for glacier melt, snow melt, and rainfall runoff do not deviate much from each other or from the respective ensemble means. Projections for total water runoff from the GISS model are about 7 percent higher than the ensemble mean and can be explored further. Consistent increases in glacier melt and decreases in snow melt across the GCMs are caused by warming temperatures, transferring heat more easily, melting glaciers, and causing more precipitation to fall as rain rather than snow.

RCP 4.5 (MITIGATION) COMPARED TO THE RCP 4.5 (STANDARD) SCENARIO

The ensemble mean decadal average water runoff in the RCP 4.5 (mitigation) scenario is 27–32 percent lower in each of the basins compared to the RCP 4.5 (standard) scenario (and also compared to the reference period) (table 6.6). These declines, when accompanied by rising demand for water, will exacerbate water stress in the entire region.

FIGURE 6.2 Total Water Runoff, Rainfall Runoff, and Water Released from Snow Melt and Ice Melt in the Himalaya, Karakoram, and Hindu Kush Region under RCP 4.5 (Standard) Scenario

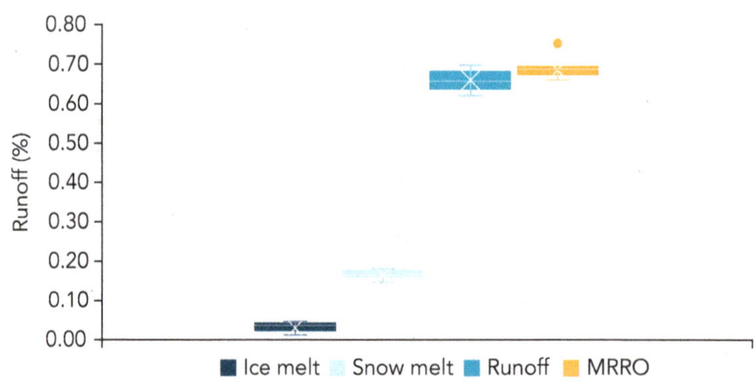

Source: Based on data from the Hadley Centre for Climate Prediction and Research regional climate model (HadRM3P).

Note: MRRO = total surface runoff.

TABLE 6.6 Water Runoff and Partitioning between Sources in the Himalaya, Karakoram, and Hindu Kush Region, by Basin (RCP 4.5 Mitigation Scenario)

Basin	Annual precipitation (millimeters)	Annual runoff (cubic kilometers per year)	Contribution to total runoff (%)		
			Glacier melt	Snow melt	Rainfall runoff
Upper Ganges	1,164	894	9	27	64
Upper Brahmaputra	1,891	1,426	4	29	67
Upper Indus	772	540	10	59	31

Source: Created for this publication based on publicly available data sets.

One driver of this reduction is the lower level of precipitation (2–9 percent) under the mitigation scenario, with corresponding decreases in runoff from rainfall. While the specific reasons for the lower precipitation levels were not explored, reductions in aerosols, which act as condensation nuclei for atmospheric moisture, can lead to less precipitation in each location.

Glacier melt contributes about the same share to total water runoff in both the standard and mitigation scenarios, representing a smaller amount of total water in absolute terms compared to the mitigation case. However, the absolute levels of glacier melt are approximately half the levels in the standard scenario, suggesting that reductions in BC emissions can be effective in reducing the rate of depletion of glaciers. Snow melt, however, is projected to decline in an absolute sense in the upper Indus but to remain at about the same levels in the upper Ganges and the upper Brahmaputra in the mitigation scenario relative to the standard scenario (map 6.5 and figure 6.3). However, given the overall lower level of water runoff, the share of water derived from snow melt will increase.

The estimates of total water runoff as well as its partitions for the mitigation scenario vary from each other and the ensemble more than those for the standard case. The ensemble means for total runoff as well as the contribution from each source are statistically significant, suggesting that BC polices in South Asia have clear measurable impacts on the future availability of water and the planning of water resource use.

Summary of Findings

A calibrated version of the conceptual cryosphere hydrology framework was used to assess the impacts of BC emission reduction polices by South Asian countries on the production of water in the mountain headwaters of the Indus, Ganges, and Brahmaputra river basins in the context of a changing climate. The study produced the following findings:

- Considerable uncertainty exists regarding runoff levels in the HKHK region even at an aggregate level for the historic period. The uncertainties only increase farther out into the future.

FIGURE 6.3 Total Water Runoff, Rainfall Runoff, and Water Released from Snow Melt and Ice Melt as a Result of Aerosols for the RCP 4.5 (Mitigation) Scenario

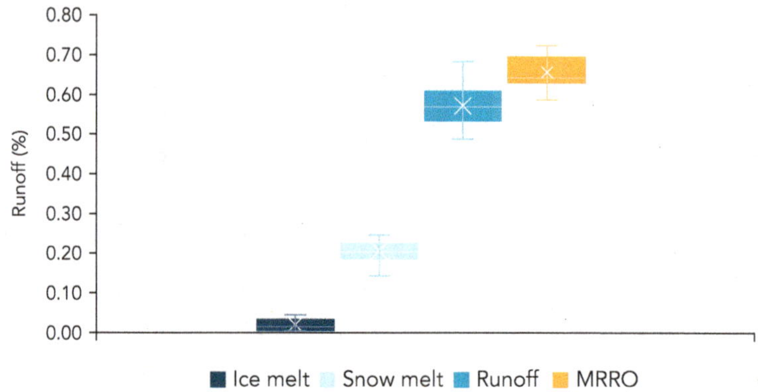

Source: Based on data from the Hadley Centre for Climate Prediction and Research regional climate model (HadRM3P).
Note: MRRO = total surface runoff.

- The long-term average temperatures in the HKHK region are projected to rise an additional 0.4°C beyond the levels already projected for an RCP 4.5 scenario if South Asian countries implement additional BC reduction policies.
- There is greater uncertainty regarding changes in the projected long-term average annual precipitation in all scenarios considered, with increases in some areas and decreases in others.
- South Asian countries can reduce BC deposition in the region 23 percent by implementing policies currently in place and an additional 50 percent by enacting and implementing new policies that are currently economically and technically feasible.
- Full implementation of current BC emissions policies in South Asia is projected to increase the water released from glacier melt both in absolute volume and as a share of total water production in the 2040s in the upstream areas of the Indus, Ganges, and Brahmaputra basins.
- Enacting additional technically and economically feasible BC emission reduction polices can reduce glacier melt to current levels, but also may reduce water supplies and exacerbate water stress.
- There is great spatial variation in these results across the region. In the context of the inherent uncertainties about the underlying data, these variations suggest that the results should be viewed as indicative of broad trends. As such, further studies focused on specific policy options are necessary prior to enacting such policies.

References

Alvarado, M. J., E. Winijkul, R. Adams-Selin, E. Hunt, C. Brodowski, C. R. Lonsdale, D. T. Shindell, et al. 2018. "Sources of Black Carbon Deposition to the Himalayan Glaciers in Current and Future Climates." *Journal of Geophysical Research: Atmospheres* 123 (14): 7482–505. doi:10.1029/2018JD029049.

Armstrong, R. L., K. Rittger, M. J. Brodzik, A. Racoviteanu, A. P. Barrett, S.-J. Singh Khalsa, B. Raup, et al. 2019. "Runoff from Glacier Ice and Seasonal Snow in High Asia: Separating Meltwater Sources in River Flow." *Regional Environmental Change* 19: 1249–61. doi:10.1007/s10113-018-1429-0.

Lutz, A. F., W. W. Immerzeel, A. B. Shrestha, and M. F. P. Bierkens. 2014. "Consistent Increase in High Asia's Runoff due to Increasing Glacier Melt and Precipitation." *Nature Climate Change* 4: 587–92.

Rangwala, I., and J. R. Miller. 2012. "Climate Change in Mountains: A Review of Elevation Dependent Warming and Its Possible Causes." *Climatic Change* 114 (3-4): 527–47. doi:10.1007/s10584-012-0419-3.

CHAPTER 7

Conclusion and Implications

Melting glaciers pose significant risks to the stability of water resources in the South Asia region, with implications for local, national, and regional economies. South Asia is highly dependent on water provided by glaciers, making the region highly vulnerable to the challenge of climate change. Future projections of climate change and black carbon (BC) for the region suggest that rising temperatures and deposition of BC will continue to affect the dynamics of snow and glaciers. While specific local impacts may be severe, glacier melt will affect the overall balance of water resources for all countries in the region.

This report set out to understand whether regional BC policies could make a difference in protecting South Asia's water towers in the context of global climate change. The report was not designed to provide definitive uncertainty bounds but rather to provide a first look into the relationship from a regional modeling perspective. To that end, the answer is "yes": BC emissions by South Asian countries play a measurable role in the availability of glaciers, snow, and water. Yet global climate change and BC emissions from outside the region also have a significant impact. This report's findings have several implications.

Implications of the Findings

Significant uncertainty remains about the basic state of water resources in the Himalaya, Karakoram, and Hindu Kush (HKHK) mountains. Comparing this report with two other reputable regional assessments reveals a lack of consensus on the basic contributions of glacier melt to runoff—and even the amount of precipitation—within the

HKHK mountain region. While previous basin-scale studies have either treated glacier dynamics adequately (Shea et al. 2015) or modeled the hydrologic changes without fully coupling the hydrologic and glacier models (Lutz et al. 2014), none has implemented a coupled glacier and hydrologic model. This report adds value and reduces uncertainty by effectively modeling snow and glacier science, BC, and hydrology with regional and global impacts. Effective water management and regional cooperation are a challenging agenda for the region given the lack of a common understanding of the current state of snow cover and glaciers. The unknowns and uncertainties posed by climate change and BC therefore require joint adaptation strategies to counteract predicted changes in these valuable environmental assets, which will have serious regional and downstream impacts if they experience heavy reductions. Potential differential impacts across HKHK basins could be handled through knowledge sharing and timely actions enhanced through regional collaboration to identify potential impacts and share best practices to mitigate them.

As observed in this report, melting glaciers, loss of seasonal snow, and changes in precipitation pose significant risks to water resources in the South Asia region and will only get worse. The findings suggest not only that climate change is threatening South Asia's water towers, but also that the impacts on glaciers and water availability may differ substantially among basins. This report shows that, in addition to the threats from global climate change, the amount of BC produced and circulated within the region not only decreases the reflectance of the glacier surface, which increases the absorption of solar radiation, but also raises air temperatures, which increases glacier melt. While it may be helpful to think of glacier meltwater as a form of insurance in the interim, especially in times of drought and dry seasons, the concern going forward is that a permanent reduction in water flow from this source will cause significant downstream effects once the region has drawn down this insurance. Given that more than 750 million people depend on the glacier and snow-fed Indus, Ganges, and Brahmaputra basins for freshwater, changes in the volume and timing of flows will have important economic and social implications.

Regional cooperation can be an effective transboundary solution, helping countries in the HKHK to manage glaciers and related natural assets collaboratively. The findings suggest that glaciers and snowpack play an important role in the timing and magnitude of seasonal streamflow within the Indus, Ganges, and Brahmaputra basins. Each basin has variations in its dependence on mountain headwaters, water demands, and ability to absorb changes in the timing of streamflow. The commonality of the BC challenges and uncertainties posed by climate change, however, suggest that the joint adoption of strategies to counteract predicted changes in snow cover and glaciers will bring many benefits, helping to avoid heavy reductions that will have serious regional and downstream impacts. Regional cooperation can also help countries to manage BC sources and natural disasters, such as glacial lake outburst floods (GLOFs), by sharing weather discharge updates to improve flood forecasting and monitoring.

Managing BC is critical to managing the region's water resources. A key message emerging from this report is that reducing atmospheric concentrations of BC has significant implications for the future of water resources in South Asia. In addition, it can result in improved public health and slow the rate of near-term climate change. The health benefits from implementing BC mitigation measures would be realized immediately. Through the historic baseline BC analysis, this report assesses the relative impact that each source (diesel engines, brick making, cookstoves fueled by biomass, open fires, kerosene wick lanterns, and agricultural practices) has on BC production and therefore on snow and glacier dynamics. Through the calculation of emission sensitivities, the report further simulates the sensitivity of BC and coemissions in conjunction with projected future climate scenarios. The findings suggest that BC not only has played a role historically in snow and glacier dynamics in the headwaters of the Indus, Ganges, and Brahmaputra basins, but also will increasingly play an important role in glacier and snow melt unless managed effectively. The report demonstrates that managing BC emissions in South Asia carries the potential to achieve global and regional climate benefits, while also capitalizing on several regional and local cobenefits. In addition, acting on BC would help to reduce regional warming and the disruption of regional weather patterns, such as the monsoonal systems.

Global climate change driven by greenhouse gases (GHGs) is also driving extensive changes to the region's water resources. Climate change may cause the supply of water to fluctuate both temporally and spatially and may alter the global hydrologic cycle. By 2050, from 1.5 billion to 1.7 billion people in South Asia (70–81 percent of the population, according to the World Resources Institute) are projected to be exposed to water scarcity. A recent World Bank report estimates that rising temperatures and changing patterns of monsoon rainfall due to climate change could lower the living standards of half the regional population by 2050 (Mani et al. 2018).

Countries of the region are taking steps to curb BC emissions through enhancing fuel efficiency standards for vehicles, phasing out diesel vehicles and promoting electric vehicles, accelerating the use of liquefied petroleum gas for cooking and other clean cookstove programs, and upgrading brick kiln technologies. In the Rohtang Pass region of India, even vehicular movement is restricted to lower the emissions of BC in proximity to glaciers and snow-covered areas. However, the report outlines a number of cost-effective measures that are available and could be put in place to curb future BC emissions.

Improving the efficiency of brick kilns could be key to managing BC. The findings of the BC analysis indicate that industry (primarily brick kilns) and residential burning of solid fuel are the main contributors of BC, accounting for 45–66 percent of the regional contribution to anthropogenic BC deposition; on-road diesel fuels make a smaller contribution (7–18 percent), and open burning accounts for less than 3 percent in all seasons. Several factors influence BC emissions from brick making, including the technology used, the source of fuel, and the way that brick kilns are operated and maintained. While mitigation options that reduce BC emissions by introducing new or modified technologies or fuels are typically more costly than traditional kilns or fuels, the number of modest up-front investments that pay off quickly now is growing. A recent World Bank report identifies a few cost-effective technology solutions geared toward cleaner brick kiln operations (Eil et al. 2020). The report suggests that transitioning to cleaner, more efficient technologies in the brick kiln sector would require government support through regulations and standards, enforcement, legislative mandates, and other enticements, such as preferential permitting, access to markets, and concessional financing. Further, implementing these emission reduction methods also would require a change in behavior (for example, how a kiln is operated or maintained).

Transition to cleaner cookstoves will have multiple benefits. The findings also suggest that residential burning of solid fuel is a major contributor to the generation of BC in the region. Transitioning to new fuels or improved stoves could significantly reduce BC deposition, while at the same time improving the health and lives of women and children. However, several programs supporting clean cookstoves have met with only limited success, in some cases due to behavioral aspects and lack of a supply chain. In other cases, difficulties have involved affordability, stove performance, design, and long-term adoption. Alternatively, BC emissions from the residential sector could be reduced by switching households to a cleaner fuel, which in many cases would mean moving from biomass or coal to kerosene or liquefied petroleum gas. In India, the central government and some state governments have launched programs to implement fuel switching among low-income households in urban and rural areas, with some success. These programs could potentially be replicated in other parts of the region.

Reducing the use of diesel fuel can have a direct impact on cutting BC deposition in the HKHK. Historically, high-income countries have successfully relied on improvements in the quality of fuel and vehicle emissions standards to control and reduce BC emissions from diesel fuel combustion. New vehicles should be equipped with diesel particulate filters that meet strict emissions standards (Euro 6/VI). However, doing so will require investing in and upgrading refineries that produce the needed ultra-low-sulfur diesel or else increasing imports to ensure sufficient supplies. Other beneficial programs include the replacement, retrofitting, inspection, and maintenance of vehicles, along with complementary policies to lower the demand for travel and the long-term growth in emissions, such as fuel taxes, congestion charges, and logistics management. The World Bank is a partner in the Climate and Clean Air Coalition, which supports strategies to reduce BC emissions from existing fleets, and the Sustainable Mobility for All partnership, which promotes reducing excess road use, expanding the share of greener and safer transport modes and creating new ones, and avoiding unnecessary trips.

BC reduction also offers a fast-action solution to slow warming. In the long run, switching to solar and other clean energy solutions can control BC flow in the region. Based on the accumulated knowledge of the behavior of short-lived climate pollutants (BC, methane, ozone) in the Himalayan atmosphere and their impacts on glacier melting, countries in the region must take urgent, immediate, and results-based measures specifically to target the emissions of BC and ozone precursors to reduce the content of these short-lived pollutants, especially during the premonsoon season. Combating BC can solve a range of important problems at once. It can reduce household pollution, saving many lives; it can protect biodiversity and improve ecosystem goods and services, including clean water, clean air, and better productivity; and, above all, it can slow down glacier melting to ensure the availability of freshwater flow in the Himalayan rivers on which close to 1.3 billion people's lives and livelihoods depend.

Managing water resources now is key to mitigating potential impacts from glacier melt. The report findings suggest that, by the middle of the century, glacier and snow melt hastened by climate change and BC will affect the supply of water through changes in the hydrologic cycle, precipitation, and temperature. If current rates of glacial retreat continue, some higher-elevation areas could experience altered water flows in the immediate near future. Eventually, shifts in the location, intensity, and variability of rain and snow due to climate change will likely have a greater impact on regional water supplies. The current inefficient allocation and use of water will further aggravate these impacts both upstream and downstream. In addition, population growth will increase the demand for water for household use, while other factors, including groundwater depletion, could have a greater impact on the region's water security overall. Changes in the availability of water resources could play an increasing role in political tensions, especially if existing water management institutions do not account better for the broader social, economic, and ecological aspects of the region. Basin-based water management to manage the water of common rivers and to cope with potential water-related problems becomes even more urgent in the South Asian context. It is also important to consider water's many values and to guide the transparent incorporation of these values into the decision making of policy makers, communities, and businesses.

Mountain communities, which are fragile in general, face additional risks from climate change, including an increasing likelihood of loss of water resources. Changes in water resources are particularly worrisome because the well-being of many mountain communities depends on a single water source, such as melt from a specific glacier or snowfield, and a single activity, such as subsistence farming. Changes in water and agricultural production in mountain areas may also have negative impacts on broader societies in South Asia. The report's findings suggest that higher temperatures and increased BC deposition will enhance glacier melt, releasing more water in the short term, until glaciers retreat significantly. Some villages may initially benefit from these effects. However, the phenomenon will also reduce the amount of water stored seasonally as snow. Eventually, snow and glaciers may both disappear or decrease substantially in many areas where people have depended on them. While some communities are already adapting to these changes, there is a need to disseminate more knowledge and build more capacity for adaptation in remote communities.

GLOFs pose serious threats. Regional projections for the lower Indus, Ganges, and Brahmaputra Rivers suggest that floods will become noticeably more frequent by the middle of the century, putting millions of people at risk. At the same time, the formation of glacier lakes in the headwaters and potentially destructive floods from the sudden emptying of these moraine-dammed lakes could harm downstream communities and infrastructure. While some recent studies have attempted to map the risks posed by these GLOFs (Maharjan et al. 2018), the increasing melt rate in coming decades will bring greater risks in proportion to the amount of additional water. The International Centre for Integrated Mountain Development (ICIMOD) and other groups are setting up early warning systems for glacial lakes. Other preventative measures could include removing loose rocks and debris that make the bursts of water even more destructive, draining glacial lakes to reduce the amount of water released by a breach and discouraging settlement in GLOF hazard zones. In the short term, it is important to identify which lakes need further risk analysis and to assist in regional climate change adaptation strategies. Appendix E provides an overview of the particular risks posed to mountain and downstream communities by flood events related to glacier melt.

The South Asia countries need to manage their hydropower and storage resources carefully. A major challenge relates to supplying enough energy to meet the aspirations of the region's growing population, especially concerning food production and other economic activities. Hydropower from the HKHK mountain systems can help to satisfy the region's growing energy needs and enhance economic prosperity through energy trade and security. Hydropower also can provide local, national, and global environmental benefits by reducing the consumption of fuelwood and fossil fuels. In addition, harnessing the huge untapped hydropower resources in the region can be an engine of economic growth as well as food security. The simulations undertaken for this report suggest that the availability of water from glaciers will depend to a large extent on the contribution of meltwater to total discharge, seasonal differences in a future characterized by climate change, future amounts of precipitation (rainfall and snowfall), and future intensity of precipitation. Hydropower developers that are still planning and building their projects need to take into account the possibility of changing water flows. With sustained or increased water flow from increased snow and glacier melt expected in the rivers until at least 2040, planned hydropower projects should be reassessed. The variation of water flows may boost the case for large water storage projects to stabilize availability over the years.

The Way Forward

This report is designed for policy makers, experts in client countries, development communities, civil society, academic researchers, journalists, and other stakeholders in South Asia. The findings are intended to inform policy makers and stakeholders and to influence public opinion at-large. For this reason, while the final report is based on state-of-the-art science, the findings will be distilled further and communicated in accessible language through a dissemination series. The overall objective is also to provide sound analytical underpinnings for policy making, investments, and other initiatives that could be supported by World Bank–financed operations. The team has deliberately focused on the scientific aspects to provide a solid understanding on the subject. It is expected that the findings will be especially useful to future task teams in their efforts to prepare hydro investments along the basin, in agriculture, or in livelihood projects or to undertake climate-water related economic and sector work at the sub-basin level. In other words, the report aims to provide a "public good" for future use by task teams who can then customize the findings to suit their specific needs. Future research could focus on quantifying the potential economic impacts of climate change on the Himalaya glaciers across countries or sectors.

The conceptual cryosphere hydrological framework (CCHF) model developed here is an open-source model. It has helped to advance the state of the art and is now available for others to use and improve on. The CCHF is easily configurable. Users can choose the process representations that best explain an area's hydrology by applying the model to specific basins or glaciated areas. The model is flexible enough that users can build their own process representations.

References

Eil, A., J. Li, P. Baral, and E. Saikawa. 2020. *Dirty Stacks, High Stakes: An Overview of Brick Sector in South Asia.* Washington, DC: World Bank.

Maharjan, S. B., P. K. Mool, W. Lizong, G. Xiao, F. Shrestha, R. B. Shrestha, N. R. Khanal, et al. 2018. *The Status of Glacial Lakes in the Hindu Kush Himalaya.* ICIMOD Research Report 2018/1, Kathmandu: International Centre for Integrated Mountain Development.

Lutz, A. F., W. W. Immerzeel, A. B. Shrestha, and M. F. P. Bierkens. 2014. "Consistent Increase in High Asia's Runoff due to Increasing Glacier Melt and Precipitation." *Nature Climate Change* 4: 587–92.

Mani, M., S. Bandyopadhyay, S. Chonabayashi, A. Markandya, and T. Mosier. 2018. *South Asia's Hotspots: The Impact of Temperature and Precipitation Changes on Living Standards.* South Asia Development Matters. Washington, DC: World Bank.

Shea, J. M., W. W. Immerzeel, P. Wagnon, C. Vincent, and S. Bajracharya. 2015. "Modelling Glacier Change in the Everest Region, Nepal Himalaya." *The Cryosphere* 9 (3): 1105–28. doi: 10.5194/tc-9-1105-2015.

APPENDIX A

Climate Model Selection and Bias Correction

This appendix discusses climate model selection and bias correction used in this research. The first section discusses the importance of climate model selection and how the specific models used in this book were selected based on the analysis contained in Mani et al. (2018). The second section discusses bias correction methodology and how bias correction was conducted for the research. The third section reviews how bias correction affects future temperature and precipitation results for the specific ensemble of models used here.

Climate Model Selection

Climate models are the primary tool for projecting how the climate will respond to changes in atmospheric greenhouse gas (GHG) emissions and what regional climate patterns will be in future decades. Climate models seek to represent relevant physical processes and linkages between solar forcing, the atmosphere, land surface, and oceans.

Choosing models that bracket the uncertainty can result in an overrepresentation of uncertainty because outlier models are often included. Selecting climate models based on their historic performance can also be problematic because correspondence between historic and projected performance is often low.

Appendix D in Mani et al. (2018) details the climate model selection process used in that study. This book uses the assembly of models selected in that earlier study, as detailed in chapter 4. Eighteen global climate models of the Coupled Model Intercomparison Project, fifth phase (CMIP5) that have publicly available data for the representative concentration pathway (RCP) 4.5 and RCP 8.5 historic periods are evaluated for Afghanistan, Bangladesh, Bhutan, India, the Maldives, Nepal, Pakistan, and Sri Lanka for three seasons: premonsoon (March–May), monsoon (June–September), and postmonsoon (October–February).

Mani et al. (2018) select 11 of the 18 climate models based on their performance in modeling historic South Asian climate. These 11 climate models were also used in this book as an ensemble to project long-term changes in average temperature and precipitation in the region for the RCP 4.5 (standard) and RCP 4.5 (mitigation) scenarios.

Climate Model Bias Correction

Climate model simulations are often bias corrected to ensure that their representation of climate is consistent with reference data sets. Many types of bias correction exist, and each method makes different sets of assumptions. A central tenet of bias correction methods is that climate models better represent the sensitivity of climate to changes in external forcings than climate corresponding to a specific set of external forcings. Stated more simply, the assumption is that climate models better represent changes in climate over time than climate at a specific time.

Two nonparametric methods tested here are empirical quantile-to-quantile (Q2Q) mapping, in which linear interpolation is used to fit values between data points (Hopson and Webster 2010) and empirical QM (eQM), in which data points are binned into 100 percentiles and nearest-neighbor interpolation is used to fit values between data points (Mosier, Hill, and Sharp 2017). Q2Q and eQM result in very similar bias corrected distributions. The comparison is conducted using daily precipitation and mean temperature, with CCSM4 (Neale et al. 2010) as the simulation and ECMWF Re-Analysis (ERA)-ERA-Interim (Dee et al. 2011) as the reference data set. During the historical period, the average bias in CCSM4 is about 150 percent for precipitation and almost 2°C for temperature. Both methods remove similar amounts of bias from the CCSM4 simulation output. The mappings generated using data during the historic baseline period are then applied to a CCSM4 RCP 4.5 simulation. Slight differences are present between the bias corrected RCP 4.5 precipitation output produced using Q2Q and eQM. For the quantiles where differences are present, the magnitude of disagreement is approximately 0.1 millimeter per day, which is a magnitude of order smaller than the bias present in the original simulation data.

Methods to Assess the Impacts of Ensemble Selection and Bias Correction

Two types of analysis are used to assess the impacts of ensemble selection and bias correction on climate projections. The first method is a time-series analysis in which climate values are aggregated over the Himalaya, Karakoram, and Hindu Kush (HKHK) region. Time-series analysis captures the temporal evolution of climate changes and associated ensemble uncertainty. These yearly time series are temporally smoothed by calculating the 30-year running mean time series for each model. The smoothed time series for individual models are then averaged to calculate the multimodel mean (MMM) for the full ensemble and CCSM4 ensemble. Confidence intervals are calculated at the 95th percentile by bootstrapping the temporally smoothed time series.

The second type of analysis compares the 30-year average MMM for the full ensemble and the CCSM4 ensemble. The 35-year average MMM captures spatial differences in climate sensitivity across the HKHK region. Where maps are depicting differences, hash marks are used to depict grid cells where the differences are not statistically distinct (for example, difference between values corresponding to RCP 4.5 and the historical baseline). The method for determining statistical distinctness is based on the method of Proistosescu, Rhines, and Huybers (2016).

ENSEMBLE SELECTION AND UNCERTAINTY IN PROJECTIONS

Climate models tend to reproduce temperatures better than precipitation (Mani et al. 2018). Despite this, the difference in temperatures between models in the full ensemble is approximately 5°C for the annual season and 3°C for the monsoon season. In contrast, the uncertainty associated with model initialization is approximately 0.2°C. Bias correction causes the cumulative distribution functions (CDFs) of all climate model simulations to match the ERA-Interim CDF during the historic baseline period. This substantially reduces the difference in the temporally smoothed climate model ensembles to around 0.2°C during the baseline period. The final result is that the MMM climate projection is generally more reliable than any single climate model.

BIAS CORRECTION AND HISTORIC SPATIAL REPRESENTATION

Temperatures in South Asia are highly diverse, with the hottest annual temperatures occurring in the Indian subcontinent and the coldest temperatures occurring in the Himalaya mountains. Without bias correction, the full ensemble and the CCM4 model ensemble differ significantly from ERA-Interim and depict markedly different climatic patterns across the region.

Bias correction practically eliminates all the differences between the ensemble MMM and ERA-Interim during the historic baseline period. This is expected and demonstrates that bias correction is effective. There is some spatial cohesion in the small differences between the climate model MMMs and ERA-Interim; however, the reason for this cohesion is not immediately apparent because bias correction is carried out independently at each spatial grid cell.

Three main areas in South Asia have significant average annual precipitation: Bangladesh and the areas east and south, the southwestern tip of India, and the southern front of the Himalaya mountains. The first two are caused by warm moist air blowing from the ocean to land. The higher precipitation along the Himalayas is orographic precipitation that results from the cooling of air as it rises to pass over the mountains.

The monsoon is the predominant feature of yearly precipitation for most locations in South Asia. The areas with the largest monsoon precipitation are in and around Bangladesh, followed by southwestern India. The wettest locations receive an average of around 20 millimeters per day, which corresponds to more than 10 meters over the four months of the monsoon season.

The full-model MMM better represents the spatial patterns of historic precipitation than the CCSM4 MMM. The largest differences between the full ensemble MMM and ERA-Interim are over Bangladesh and the surrounding areas. The CCSM4 ensemble MMM contains these differences and overrepresents the orographic precipitation along the Himalaya mountains. For both ensembles, bias correction significantly reduces differences between the climate model MMM and ERA-Interim.

ENSEMBLE SELECTION AND THE SPATIAL PATTERN OF PROJECTIONS

The magnitude of temperature increases predicted by the bias corrected full ensemble and CCSM4 ensemble are relatively similar across South Asia. In general, the full ensemble projects slightly larger temperature increases than the CCSM4 ensemble. In the HKHK region, these differences are about 0.5°C during 2030–59 and 0.75°C during 2070–99. For all locations, the differences between temperature projections for the two ensembles are statistically indistinguishable, meaning that the confidence intervals overlap.

Projected changes in precipitation are relatively small for most locations in South Asia. The largest projected change is present in the CCSM4 MMM over the eastern tip of the Himalaya mountains. This feature is present for the annual and monsoon seasons and for both projection time periods. As with temperature, the differences between climate ensembles are not statistically distinct for any locations.

CCSM4 PRECIPITATION PROJECTIONS AND THE AVERAGE FOR THE FULL ENSEMBLE

All the climate models from both ensembles project that temperatures will continue rising, but trends in precipitation are more ambiguous. In particular, Lutz, Immerzeel, and Kraaijenbrink (2014) find that some climate models project precipitation increases and others project precipitation decreases in the upper Indus basin. The total range of increases under RCP 4.5 and RCP 8.5 is −10 to +17 percent for 2021–50 relative to 1961–90. Their analysis is based on all climate models participating in CMIP5 for which outputs are available. It is well established that climate models are more uncertain regarding changes in precipitation than changes in temperature (Mani et al. 2018).

For the HKHK region, the climate models selected for both the full ensemble and the CCSM4 ensemble project that both temperature and precipitation will increase during the annual and monsoon seasons. The magnitude of changes varies by climate, model, scenario, and time frame, but the range of projected changes in annual precipitation based on bias corrected climate model inputs under RCP 4.5 is approximately 0–11 percent for 2030–59 and 1–15 percent for 2070–99. Under RCP 8.5, the projected changes in annual precipitation are approximately 1–11 percent for 2030–59 and 6–23 percent for 2070–99. Overall, changes during the monsoon season are similar to changes during the annual season.

The CCSM4 ensemble projects changes in precipitation that are close to the median of the full ensemble. The range of precipitation projections within the CCSM4 ensemble is approximately 20 percent of the full ensemble's range. Therefore, the main difference between the precipitation projections of the two ensembles is the uncertainty.

Temperature projections by the CCSM4 ensemble are lower than the median for the full ensemble. Of the eight climate models other than CCSM4 in the full ensemble, only one or two project lower temperature increases than CCSM4, depending on the scenario, time frame, and season. Generally, the three realizations of CCSM4 have similar sensitivity to temperature, which indicates that temperature is less sensitive to model instance than precipitation. Stated another way, it appears that temperature response is more fundamental than precipitation response to a given climate model.

Bias correction modifies the changes projected by individual climate models and the overall range of ensemble projections. Bias correction tends to lower the range of projected temperature increases; this also appears to be true for each individual climate model. Two drivers of this observation are that climate models tend to project larger temperature ranges than observations and are more sensitive to GHG emissions than observations.

Bias correction modifies projections of precipitation change in less predictable ways. In many instances, bias correction does not modify the overall range of the ensemble, but it does modify the projections of individual climate models within the ensemble. For example, the ensemble range for projected changes in annual precipitation from 2030 through 2059 under RCP 4.5 is approximately 1–11 percent increases for both the original and bias corrected full ensemble. Yet the original IPSL model projects increases of approximately 1 percent and the bias corrected model projects increases of approximately 4 percent. This general finding is a symptom of the uncertainty present in climate model representations.

References

Dee, D. P., S. M. Uppala, A. J. Simmons, P. Berrisford, P. Poli, S. Kobayashi, U. Andrae, M. A. Balmaseda, G. Balsamo, and P. Bauer. 2011. "The ERA-Interim Reanalysis: Configuration and Performance of the Data Assimilation System." *Quarterly Journal of the Royal Meteorological Society* 137 (656): 553–97.

Hopson, T. M., and P. J. Webster. 2010. "A 1–10-Day Ensemble Forecasting Scheme for the Major River Basins of Bangladesh: Forecasting Severe Floods of 2003–07." *Journal of Hydrometeorology* 11 (3): 618–41. https://doi.org/10.1175/2009JHM1006.1.

Lutz, A. F., W. W. Immerzeel, and P. Kraaijenbrink. 2014. "Gridded Meteorological Datasets and Hydrological Modelling in the Upper Indus Basin: Final Report." FutureWater Report 130, FutureWater, Wageningen.

Mani, M., S. Bandyopadhyay, S. Chonabayashi, A. Markandya, and T. Mosier. 2018. *South Asia's Hotspots: The Impact of Temperature and Precipitation Changes on Living Standards.* South Asia Development Matters. Washington, DC: World Bank.

Mosier, T., D. F. Hill, and K. V. Sharp. 2017. "Update to the Global Climate Data Package: Analysis of Empirical Bias Correction Methods in the Context of Producing Very High-Resolution Climate Projections." *International Journal of Climatology* 38 (2): 825–40. doi: 10.1002/joc.5213.

Neale, R. B., A. Gettelman, S. Park, C.-C. Chen, P. H. Lauritzen, D. L. Williamson, A. J. Conley, et al. 2010. "Description of the NCAR Community Atmosphere Model (CAM 5.0)." NCAR Technical Note NCAR/TN-486+ STR, University Corporation for Atmospheric Research, Boulder, CO. https://www.ccsm.ucar.edu/models/ccsm4.0/cam/docs/description/cam4_desc.pdf.

Proistosescu, C., A. Rhines, and P. Huybers. 2016. "Identification and Interpretation of Nonnormality in Atmospheric Time Series." *Geophysical Research Letters* 43 (10): 5425–34. https://doi.org/10.1002/2016GL068880.

APPENDIX B

Black Carbon Transport and Impacts

This appendix provides an overview of the aerosol and black carbon (BC) modeling conducted for this book. Two climate models were utilized. The first is the GISS model E2, a global climate (general circulation) model developed by the Goddard Institute for Space Studies (GISS) to develop custom climate simulations based on a standard and a low-carbon scenario. Second, to develop a more detailed understanding of impacts, the higher-resolution Weather Research and Forecasting (WRF)-Chem model (Alvarado et al. 2018) is used, using outputs from the GISS modeling exercise. The complete methodology and results of Alvarado et al. (2018) are available in the published version of that paper and in chapter 5 of this book.

Full discussion of methodology and results, as well as supporting charts and figures, are omitted because of space concerns but are available upon request.

Methodology

Two sets of experiments are run with the GISS-E2-R climate model (Schmidt et al. 2014). The first is consistent with representative concentration pathway (RCP) 4.5 associated with the Coupled Model Intercomparison Project, fifth phase (CMIP5) (Taylor, Stouffer, and Meehl 2012). The second is consistent with representative concentration pathway (RCP) 4.5, except that aerosol emissions are reduced for countries in South Asia. This appendix compares the results of these two experiments to each other and to a simulation of historic climate using monthly precipitation, mean temperature, and temperature range. Future scenarios are analyzed for two periods between 2030 and

2099 by examining their respective impacts on long-term average climate, distribution of weather, and relationship to El Niño-Southern Oscillation, or ENSO for short.

GISS model simulations of the two emissions scenarios for future years are conducted for 2006–2100. In places, these simulations are compared to a single simulation of the historic baseline period (1970–1999).

Analyses of future years are conducted using GISS-E2-R model (Schmidt et al. 2014) output for 2030–59 and 2070–99: "mid-century" and "end-of-century," respectively. For each of these periods, output is analyzed for the monsoon season (June–September) and then for the entire year.

Daily output is aggregated to a monthly time step to increase the available number of ensemble members. Units of precipitation used for both time steps are millimeters per day, so aggregation simply involves calculating the mean for all days in a month. For temperature range, daily minimum and maximum temperatures are first aggregated to the monthly time step. The monthly temperature range is the difference in the monthly mean maximum and monthly mean minimum temperatures.

EMISSIONS EXPERIMENTS

Two sets of simulations are conducted to assess the impacts of aerosols emitted in South Asia for 2006 through 2100. The first scenario, referred to as RCP 4.5 (st), exactly follows the RCP 4.5 emissions trajectories for greenhouse gas (GHG) and aerosol emissions.[1] The second scenario, referred to as RCP 4.5 (mi), follows RCP 4.5 emissions trajectories for GHG and aerosol emissions, except that aerosol emissions in South Asia are set to the ECLIPSE mitigation scenario (Stohl et al. 2015), with modifications as described in Alvarado et al. (2018).

Three ensemble members are produced for each set of future conditions from 2006 to 2100. In some instances, simulations for future years are also compared to a historical simulation using the same GISS-E2-R model configuration.

GISS MODEL CONFIGURATION

The GISS-E2-R simulations are performed using Physics Option 3, which includes prognostic aerosols and aerosol-cloud interactions with cloud microphysics. This model configuration is "Group 1"—as defined by Wang (2015)—which means that the model dynamically predicts aerosol concentrations and radiation impacts using physically explicit equations. Three future ensemble members and one GISS historic initialization are used.

ANALYSES

Simulations of precipitation and mean temperature are compared using a monthly time step. Three properties of climate are analyzed: long-term mean, distribution of weather, and relationship to ENSO. The long-term mean is the most common climate metric, but distribution of weather is analyzed to assess whether aerosols affect aspects of the distribution other than the mean. Also, ENSO is known to be correlated with the strength of the monsoon (Kumar 1999; Torrence and Webster 1999; Webster and Yang 1992).

LONG-TERM AVERAGE CLIMATE

Long-term average temperature and precipitation are the long-term average for all ensemble members of weather values for a given parameter over the season and time period being investigated. Confidence intervals for the long-term average are formulated by (a) detrending annual mean weather; (b) calculating the stationary bootstrap of values for a randomly determined subset of years and a randomly determined ensemble member (Politis and Romano 1994; Politis and White 2004); (c) finding the 5th and 95th percentiles of 10,000 bootstrap iterations, and (d) adding confidence intervals based on detrended data to the long-term mean. Differences between two scenarios are considered robust when confidence intervals do not overlap.

DISTRIBUTION OF WEATHER

Weather distribution is estimated by calculating the empirical cumulative distribution function (CDF) of monthly time series for each season and time period (100 bins). The scenarios are then compared through a six-step process (Proistosescu, Rhines, and Huybers 2016): (1) calculate the fast Fourier transform of the two time series; (2) randomize (separately) the phase of each component; (3) calculate the two-distribution Kolmogorov-Smirnov (KS) test; (4) repeat steps 1–3 10,000 times; (5) find the critical KS value, defined as the 95th percentile of the 10,000 KS values; and (6) perform the two-distribution KS test on the original two time series. The null hypothesis of identical distributions is rejected if the KS value from step 6 is greater than or equal to the critical KS value from step 5.

RELATIONSHIP TO ENSO

ENSO strength is calculated following the method of Trenberth (1997) for the Nino3.4 region. Annual mean ENSO strength is then used in ordinary least squares (OLS) linear regression to predict weather anomalies for variables, seasons, and time periods. For temperature, the anomalies are calculated arithmetically; for precipitation, they are calculated multiplicatively.

The relationship between ENSO yearly excursions and yearly weather is considered significant when the magnitude of the relationship is larger than the standard error estimated by OLS regression. Differences in the relationships between scenarios are considered confident when their confidence intervals do not overlap.

Results

LONG-TERM AVERAGE CLIMATE

The GISS model projects increased long-term average temperatures under RCP 4.5 (st) compared to the historic baseline, larger in the end-of-century than mid-century. Projected changes in long-term average precipitation contain some spatial heterogeneity, but also consistency between seasons and time horizons. Long-term average monthly temperature ranges are projected to decrease for most of South Asia.

Long-term average temperatures differ between the RCP 4.5 (st) and RCP 4.5 (mi) scenarios, but differences are smaller than the bounds of the confidence intervals. By the end-of-century period, temperature differences over South Asia are more mixed between the scenarios, with both increases and decreases. The long-term average precipitation is very similar between the RCP 4.5 (st) and RCP 4.5 (mi) scenarios for future years. The long-term average monthly temperature range is very similar overall between the RCP 4.5 (st) and RCP 4.5 (mi) scenarios.

DISTRIBUTION OF WEATHER

The characteristics of temperature, precipitation, and monthly temperature distributions are projected to change under RCP 4.5 (st) relative to the historic baseline for many areas.

Differences in temperature, precipitation, and monthly temperatures are not statistically distinct between the RCP 4.5 (st) and RCP 4.5 (mi) scenarios. This does not necessarily imply that aerosols have no effect. Instead, any differences are smaller than, or comparable in magnitude to, the differences between models in the ensemble.

Weather distributions are calculated for six locations in South Asia. Using New Delhi as an example, the distributions of temperature and monsoon precipitation under both RCP 4.5 scenarios are mostly shifted to hotter temperatures and larger precipitation events compared to the historic baseline scenario. The model projects that precipitation will increase at most percentiles of the cumulative distribution function, especially at the high end of the distribution and for the monsoon season. Monthly temperature ranges are projected to contract at all percentiles under both RCP 4.5 scenarios as compared to the historic baseline simulation.

RELATIONSHIP TO ENSO

It varies by season, but positive average ENSO excursions correspond to positive temperature anomalies across a large swath of Asia. The relationship between ENSO excursions and precipitation is less spatially coherent, with both positive and negative relationships in South Asia during the annual season. Temperature range does not have a regional-scale consistent relationship to ENSO, and the empirical relationship is not confident for most of South Asia.

In South Asia, changes in the yearly signal are not statistically significant; however, during the monsoon season, the changes are significant for the end-of-century projections, indicating a strong positive relationship between ENSO excursions and temperature anomalies. Changes in the relationship between ENSO and precipitation anomalies under RCP 4.5 have less spatial and temporal coherence and confidence relative to temperature. Overall, changes in the relationship between ENSO and temperature range between the RCP 4.5 (st) and historic baseline scenarios are not spatially coherent or confident.

Lowering aerosol emissions in South Asia has a measurable impact on the relationship between ENSO and temperature anomalies. Positive ENSO excursions are correlated with negative annual temperature anomalies in North Asia, but the differences between the two scenarios are not statistically significant for most parts of South Asia. Differences in the relationship between ENSO and precipitation under the two scenarios are largely not statistically significant. Differences in the relationship between ENSO and temperature range anomalies under the two scenarios are largely not statistically significant. This is despite the sizable and spatially coherent differences in the relationship between ENSO and mean temperature anomalies. While ENSO is a large factor in temperature excursions, temperature range excursions are driven by other processes.

Summary

Alvarado et al. (2018) detail the use of the WRF-Chem model for impacts across scenarios. The complete methodology and results are available in the published version of that paper and in chapter 5 of this book. This section summarizes the paper's key points, methodology, and results. Overall, the following are the key points of the paper:

- Sources outside South Asia have an impact on BC deposition to glaciers similar to that of South Asian anthropogenic sources.
- If emissions are mitigated in South Asia (but not elsewhere), the contribution of South Asian anthropogenic sources will decrease between 11 percent and 34 percent of the total.
- The changes due to phases of ENSO are generally small and do not follow a consistent pattern.

Alvarado et al. (2018) use the WRF-Chem model together with a modified version of the ECLIPSE 5a emissions inventory model to assess the sources of deposition to the Himalaya, Karakoram, and Hindu Kush (HKHK) region. They model deposition to the HKHK region under current conditions and for the period 2040–50 under two emissions scenarios and three phases of the ENSO cycle. The paper concludes that, under current conditions, sources within and outside have relatively similar impact to the HKHK. Specifically, sources outside of South Asia contribute between 35 percent and 87 percent of BC deposition, depending on the month, while sources within South Asia contribute between 13 percent and 62 percent. The primary sources within South Asia are brick kilns and residential solid fuel burning, accounting for 45 percent to 66 percent of deposition in the HKHK region.

With no further policy changes or controls on emissions, in the 2040–50 period, sources of deposition in the HKHK region will be drawn more heavily from South Asian sources within the study domain (45 percent to 65 percent) than from sources outside the domain (26 percent to 52 percent). If emissions are controlled or mitigated through policy decisions, the relative contribution from South Asian sources to deposition in the HKHK will decrease to 11–34 percent. Consistent patterns for the ENSO are not detectable and are an area for future work.

Note

1. See RCP Database (version 2.0), http://tntcat.iiasa.ac.at/RcpDb/dsd?Action=htmlpage&page=about.

References

Alvarado, M. J., E. Winijkul, R. Adams-Selin, E. Hunt, C. Brodowski, C. R. Lonsdale, D. T. Shindell, et al. 2018. "Sources of Black Carbon Deposition to the Himalayan Glaciers in Current and Future Climates." *Journal of Geophysical Research: Atmospheres* 123 (14): 7482–505. doi:10.1029/2018JD029049.

Kumar, K. K. 1999. "On the Weakening Relationship between the Indian Monsoon and ENSO." *Science* 284 (5423): 2156–59. doi:10.1126/science.284.5423.2156.

Mani, M., S. Bandyopadhyay, S. Chonabayashi, A. Markandya, and T. Mosier. 2018. *South Asia's Hotspots: The Impact of Temperature and Precipitation Changes on Living Standards*. South Asia Development Matters. Washington, DC: World Bank.

Politis, D., and J. Romano. 1994. "The Stationary Bootstrap." *Journal of the American Statistical Association* 89 (428): 1303–13.

Politis, D. N., and H. White. 2004. "Automatic Block-Length Selection for the Dependent Bootstrap." *Econometric Reviews* 23 (1): 53–70. doi:10.1081/ETC-120028836.

Proistosescu, C., A. Rhines, and P. Huybers. 2016. "Identification and Interpretation of Nonnormality in Atmospheric Time Series." *Geophysical Research Letters* 43 (10): 5425–34. doi:10.1002/2016GL068880.

Schmidt, G. A., M. Kelley, L. Nazarenko, R. Ruedy, G. L. Russell, I. Aleinov, M. Bauer, et al. 2014. "Configuration and Assessment of the GISS Model E2 Contributions to the CMIP5 Archive." *Journal of Advances in Modeling Earth Systems* 6 (1): 141–84. doi:10.1002/2013MS000265.

Stohl, A., B. Aamaas, M. Amann, L. H. Baker, N. Bellouin, T. K. Berntsen, O. Boucher, et al. 2015. "Evaluating the Climate and Air Quality Impacts of Short-Lived Pollutants." *Atmospheric Chemistry and Physics* 15 (18): 10529–66.

Taylor, K. E., R. J. Stouffer, and G. A. Meehl. 2012. "An Overview of CMIP5 and the Experiment Design." *Bulletin of the American Meteorological Society* 93 (4): 485–98. doi:10.1175/BAMS-D-11-00094.1.

Torrence, C., and P. J. Webster. 1999. "Interdecadal Changes in the ENSO–Monsoon System." *Journal of Climate* 12 (8): 2679–90. doi:10.1175/1520-0442(1999)012<2679:ICIT EM>2.0.CO;2.

Wang, C. 2015. "Anthropogenic Aerosols and the Distribution of Past Large-Scale Precipitation Change." *Geophysical Research Letters* 42 (24): 10876–84. doi:10.1002/2015GL066416.

Webster, P. J., and S. Yang. 1992. "Monsoon and ENSO: Selectively Interactive Systems." *Quarterly Journal of the Royal Meteorological Society* 118 (507, pt. B): 877–926. doi:10.1002/qj.49711850705.

APPENDIX C

Downscaling Climate

This appendix provides an overview of methods of downscaling climate considered and adopted for this report. The Climate Forecast Applications Network (CFAN) developed both an interim and a final report for the World Bank. This appendix primarily draws on that final report, which took results from the interim report and developed the final downscaling methodology implemented in this book. It also discusses work comparing the downscaling method selected and other available and potential methods.

Both reports, as well as supporting charts and figures, are available upon request.

CFAN Downscaling Methodology

Based on the systematic linear relationships observed between spatial gradients of precipitable water and elevation, CFAN developed the following method for precipitation downscaling:

1. By calendar month, determine the overall regional relationship between local ERA-Interim (ERAI) dPW%(x,y) and dZ(x,y) by linear regression, dPW%(x,y), from actively precipitating daily grid cells. (dPW precipitable water gradients, dZ is change in elevation.)
2. Reorient local monthly dPW% gradients to reflect the magnitude of PW% loss in the local upslope direction. This trigonometric calculation isolates local changes in atmospheric water content due to upslope cooling, typically a small adjustment. Monthly coefficients allow seasonal changes to be represented.
3. Calculate monthly mean gradients of dPW%/dZ in local upslope directions, retaining the regression coefficients for downscaling. Interpolate monthly mean 0.75°

gradients of dPW(x,y) and dZ(x,y) to 1-kilometer resolution. Calculate local interpolated (smooth) slope magnitudes in the direction of local PW gradients.

4. Interpolate 0.75° ERAI elevations to 1 kilometer and subtract these large-scale, smooth-terrain features from high-resolution 1-kilometer elevations to obtain fine-scale features. For each calendar month, calculate fine-scale slopes along local vectors of PW decline.
5. Obtain the daily 0.75 ERAI precipitation field and interpolate to 1-kilometer resolution.
6. Convert fine-scale slopes to percent precipitation adjustments, based on the appropriate monthly rate of dPW%/dZ (regression coefficient).
7. Add this adjustment to the smooth interpolated precipitation field to obtain a final downscaled precipitation field.

This new method exploits the systematic linear relationships observed between local gradients of atmospheric moisture loss (percent) and local slopes (kilometers), interpreted as an indirect measure of local orographic forcing, largely by upslope cooling. The current approach emphasizes the local role of slopes, the primary topographic influence on precipitation along the Himalaya front (Anders et al. 2006). On small spatial scales (< 0.75°), local slope effects likely dominate the cumulative depletion effects of upwind slopes. CFAN experimented with models of rain shadowing by upwind slopes and found that these cumulative effects are generally well represented by ERAI data at 0.75°.

Comparison of Downscaling Methods

The use of downscaling is controversial in many instances (Hall 2014; Hewitson et al. 2014). This section compares five sets of daily time-series inputs for four catchments in the upper Indus basin (table C.1). The data sets are assessed by using them as input to an instantiation of the conceptual cryosphere hydrology framework (CCHF) (Mosier, Hill, and Sharp 2016) and comparing each model's output to observations for glacier mass balance (MB), snow-covered area, and streamflow.

ASSESSMENT METHOD

The assessment method focuses on performance in high-elevation mountain hydrology studies in the Himalaya mountains. CCHF outputs are compared to Moderate Resolution Imaging Spectroradiometer (MODIS) observations of snow-covered area (SCA) and streamflow in all four catchments and high-resolution glacier MB observations in one catchment. Climate products are compared by using each of them to force an instantiation of the CCHF snow and glacier hydrology model. The analysis is conducted from February 2001 through October 2007, because data for all five climate products are available in those years.

TABLE C.1 Comparison of Five High-Resolution Daily Climate Products

Short name	Long name	Reference
HAR	High Asia Refined analysis	Maussion et al. 2014
IGM	Inverse glacier modeling	Immerzeel et al. 2015
ERA (p = i, t = i)	ERA-Interim, interpolated for precipitation and temperature	Dee et al. 2011
ERA (p = i, t = OR)	ERA-Interim, interpolated for precipitation and orographic relationships for temperature	Dee et al. 2011
ERA (p = OR, t = OR)	ERA-Interim, orographic relationships for precipitation and temperature	Dee et al. 2011

Sources: Dee et al. 2011; Immerzeel et al. 2015; Maussion et al. 2014.

Assessment Catchments

Four catchments in the upper Indus basin are used for assessment: Donyian, Hunza, Karora, and Naltar. Daily stream flow records are available at all four of these stations (Tahir et al. 2011). All four are relatively high elevation and are fully snow covered in an average winter (Hall, Salomonson, and Riggs 2006).

Hunza and Karora are used for CCHF calibration because geodetic MB observations are available for Hunza, and Hunza and Karora have contrasting climatic regimes due to differences in their elevation, which allows CCHF calibration parameters to account for a wider range of conditions. Donyian and Naltar are used for CCHF validation.

Observation Data

Three types of observation data are used in this assessment: glacier MB, snow-covered area (SCA), and streamflow. Comparison of model estimates and observations for each of these types of data provides indirect evidence of how physically representative the five climate products assessed are for the study region.

CCHF Instantiation

This analysis uses CCHF Version 2.0 (Mosier 2018), the same structure as Version 1 (Mosier 2016) with an updated delineation of process representations. This allows one to represent debris-covered glaciers better and improves computational efficiency. The CCHF instantiation used here calculates a Silt Density Index (SDI) factor that varies between snow and ice. The debris factor for ice is then modified based on the empirical relationship between debris cover and melt (extracted from "averaged curves" line in Kraaijenbrink et al. 2017).

CCHF is calibrated in two stages, with snow and glacier processes calibrated in the first stage and runoff and all other processes calibrated in the second stage. Glacier MB and SCA observations are used for calibration in stage 1, and streamflow is used for calibration in stage 2.

Performance of modeled glacier MB is assessed using "mass balance efficiency" (MBE), defined here as follows:

$$\text{MBE} = \sqrt{(r-1)^2 + \left((\bar{\text{m}} - \bar{\text{o}}) / \sigma_0^{\ 2}\right)} \tag{C.1}$$

where r is the Pearson Moment correlation coefficient, $\bar{\text{m}}$ is the spatially averaged change in modeled glacier mass, $\bar{\text{o}}$ is the spatially averaged change in observed glacier mass, and σ is the standard deviation in observed changes in glacier mass. The second term inside of the square root in equation C.1 is the spatially integrated bias in modeled MB scaled by the observed standard deviation.

RESULTS AND DISCUSSION

Calibration Performance

CCHF performance during calibration varies considerably between climate products (table C.2). Inverse glacier modeling (IGM) and ERA (p = i, t = OR) tie for best ability to reproduce glacier MB in Hunza. ERA (p = i, t = OR) reproduces SCA averaged between calibration sites the best, followed closely by High Asia Refined (HAR) and ERA (p = OR, t = OR). IGM reproduces SCA averaged between calibration sites the worst. HAR reproduces streamflow the best (Kling-Gupta Efficiency [KGE] = 0.70), with ERA (p = i, t = i) (KGE = 0.59) and IGM (KGE = 0.58) ranked second and third, respectively.

The calibration process seeks to minimize error averaged between study sites, which tends to reduce the differences in performance between multiple sites. Yet the differences in performance between sites vary substantially between climate products

TABLE C.2 CCHF Performance during Calibration for Each Climate Product

	Karora		Hunza			Mean		Difference	
Climate product	SCA	Flow	MB	SCA	Flow	SCA	Flow	SCA	Flow
HAR	0.05	0.62	0.26	0.43	0.58	0.24	0.60	0.38	0.04
IGM	−0.30	0.67	0.54	−0.05	0.78	−0.18	0.73	0.25	0.11
ERA (p = i, t = i)	−0.19	0.51	−0.06	0.35	0.74	0.08	0.63	0.54	0.23
ERA (p = i, t = OR)	0.08	0.46	0.59	0.41	0.64	0.25	0.55	0.33	0.18
ERA (p = OR, t = OR)	0.08	0.46	0.57	0.39	0.64	0.24	0.55	0.31	0.18

Source: Original calculations for this publication based on publicly available climate data sets.
Note: In all instances, zero model error corresponds to a value of 1. "mean" and "difference" refer to comparison of model performance between catchments. CCHF = conceptual cryosphere hydrology framework. ERA = ECMWF Re-Analysis. HAR = High Asia Refined. IGM = inverse glacier modeling. MB = mass balance. OR = orographic relationships. SCA = snow-covered area.

(table C.2). ERA (p = OR, t = OR), ERA (p = i, t = OR), and HAR have the lowest differences in performance between study sites for SCA (Parajka and Blöschl [PB] scores of 0.45, 0.46, and 0.49, respectively).

For streamflow, ERA (p = OR, t = OR), IGM, ERA (p = i, t = OR), and HAR all have quite low differences in streamflow performance between sites (KGE values vary between 0.00 and 0.06). The smaller differences in performance for streamflow compared to SCA may reflect (a) differences in sensitivity between the PB and KGE and (b) the fact that stage 1 of the calibration procedure also optimized glacier MB in the Hunza catchment.

Validation Performance

Overall patterns in performance during validation are similar to those during calibration (compare table C.3 to table C.2). ERA (p = i, t = OR) has the best SCA performance, and HAR has the best streamflow performance, in both cases when averaged between the two validation sites. This ranking is the same as during calibration. ERA (p = OR, t = OR) and HAR perform as well as ERA (p = i, t = OR) for SCA (PB values, respectively, of 0.35 and 0.32 compared to 0.37). Like calibration, ERA (p = i, t = i) is the second best-performing climate product for streamflow during validation.

Validation performance for many of the climate products is higher than calibration performance (compare table C.3 to table C.2). These results are unexpected because models typically perform best for calibration. The reason that model performance may be more similar during validation is that the two validation sites are more similar to each other than the two calibration sites. The current CCHF instantiations is a variant of a simple degree-index model, and performance of this type of model is known to vary across climatic and geographic conditions (Mosier, Hill, and Sharp 2016).

TABLE C.3 CCHF Performance during Validation for Each Climate Product

	Donyian		Naltar		Mean		Difference	
Climate product	**SCA**	**Flow**	**SCA**	**Flow**	**SCA**	**Flow**	**SCA**	**Flow**
HAR	0.16	0.57	0.41	0.66	0.29	0.62	0.25	0.09
IGM	0.29	0.49	−0.15	0.68	0.07	0.59	0.44	0.19
ERA (p = i, t = i)	0.07	0.46	0.31	0.38	0.19	0.42	0.24	0.08
ERA (p = i, t = OR)	0.35	0.64	0.52	0.73	0.44	0.69	0.17	0.09
ERA (p = OR, t = OR)	0.34	0.63	0.48	0.73	0.41	0.68	0.14	0.10

Source: Original calculations for this publication based on publicly available climate data sets.
Note: In all instances, zero model error corresponds to a value of 1. "mean" and "difference" refer to a comparison of model performance between catchments. CCHF = conceptual cryosphere hydrology framework. ERA = ECMWF Re-Analysis. HAR = High Asia Refined. IGM = inverse glacier modeling. OR = orographic relationships. SCA = snow-covered area.

References

Anders, A. M., G. Roe, B. Hallet, D. R. Montgomery, N. Finnegan, and J. Putkonen. 2006. "Spatial Patterns of Precipitation and Topography in the Himalaya." *Geological Society of America Special Papers* 398: 39–53.

Dee, D. P., S. M. Uppala, A. J. Simmons, P. Berrisford, P. Poli, S. Kobayashi, U. Andrae, M. A. Balmaseda, G. Balsamo, and P. Bauer. 2011. "The ERA-Interim Reanalysis: Configuration and Performance of the Data Assimilation System." *Quarterly Journal of the Royal Meteorological Society* 137 (656): 553–97.

Hall, A. 2014. "Projecting Regional Change." *Science* 346 (6216). doi:10.1126/science.aaa0629.

Hall, D. K., V. V. Salomonson, and G. A. Riggs. 2006. "Modis/Terra Snow Cover Monthly l3 Global 0.05degcmg, Version 5, Mod10a2." Technical report, NSIDC (National Snow and Ice Data Center), Boulder, CO.

Hewitson, B. C., J. Daron, R. G. Crane, M. F. Zermoglio, and C. Jack. 2014. "Interrogating Empirical Statistical Downscaling." *Climatic Change* 122 (4): 539–54. doi:10.1007/s10584-013-1021-z.

Immerzeel, W. W., N. Wanders, A. F. Lutz, J. M. Shea, and M. F. P. Bierkens. 2015. "Reconciling High Altitude Precipitation in the Upper Indus Basin with Glacier Mass Balances and Runoff." *Hydrology and Earth System Sciences* 19 (11):4673–87. doi:10.5194/hess-19-4673 2015.

Kraaijenbrink, P. D. A., M. F. P. Bierkens, A. F. Lutz, and W. W. Immerzeel. 2017. "Impact of a Global Temperature Rise of 1.5 Degrees Celsius on Asia's Glaciers." *Nature* 549 (7671): 257–60. doi:10.1038/nature23878.

Maussion, F., D. Scherer, T. Mölg, E. Collier, J. Curio, and R. Finkelnburg. 2014. "Precipitation Seasonality and Variability over the Tibetan Plateau as Resolved by the High Asia Reanalysis." *Journal of Climate* 27 (5): 1910–27. doi:10.1175/JCLI-D-13-00282.1.

Mosier, T. 2016. thomasmosier/CCHF: Initial Public Release.

Mosier, T. 2018. thomasmosier/CCHF: Version 2.0.

Mosier, T., D. F. Hill, and K. V. Sharp. 2016. "How Much Cryosphere Model Complexity Is Just Right? Exploration Using the Conceptual Cryosphere Hydrology Framework." *The Cryosphere* 10 (5): 2147. doi:10.5194/tc-10-2147-2016.

Tahir, A. A., P. Chevallier, Y. Arnaud, L. Neppel, and B. Ahmad. 2011. "Modeling Snow Melt-Runoff under Climate Scenarios in the Hunza River Basin, Karakoram Range, Northern Pakistan." *Journal of Hydrology* 409 (1–2): 104–17.

APPENDIX D

CCHF Calibration and Validation

This research used the conceptual cryosphere hydrology framework (CCHF) to model glacier hydrology processes and simulate changes in water production in the Himalaya, Karakoram, and Hindu Kush (HKHK) region under changing climate and policy changes to reduce black carbon emissions in South Asia. CCHF is a flexible tool with nine process modules representing the key processes in the cryosphere. Its modular design facilitates easy interchange of specific representations with alternate representations. The CCHF also has built-in functionality and metrics to evaluate model skill in replicating historic observations on stream flow, ice mass balance, and snow cover. The model used for the simulations was validated in the calibration sites for periods prior to those used for the calibration as well as on the entire HKHK region. This appendix presents the final model used in the simulations.

CCHF can be implemented in calibration, validation, and simulation modes. The calibration mode allows users to input observation data and choose a calibration method (Monte Carlo, genetic algorithm, linear parameter space exploration, particle swarm optimization, or a combination thereof). In this instance, calibration began with 2,000 parameter sets based on Monte Carlo simulations to explore the parameter space and then proceeded with adaptive particle swarm optimization (Zhan et al. 2009) until convergence was reached. Validation mode requires observations and uses outputs from the set of parameters from a successful calibration run to assess model performance. Simulation mode is used for a calibrated model without requiring observations and therefore is used to project future conditions. CCHF accepts several types of observation data, including streamflow, snow and glacier stake measurements, remotely sensed snow- and ice-covered area, and gridded geodetic glacier mass balances. Observation data is used for streamflow, remotely sensed snow- and ice-covered area, and gridded geodetic glacier mass balances.

The CCHF's modular structure, when combined with its built-in calibration functions, facilitates straightforward scalable application to small or large regions. The benefit of the flexible process representations is that the user can assess which combinations of process representations are best suited for a given application of the model. For example, some combinations of processes may represent heterogeneity better across a diverse region. CCHF has previously been applied in Alaska (Mosier, Hill, and Sharp 2016), Oregon (Mosier, Hill, and Sharp 2016), and the upper Indus basin (Qian et al. 2015). Alaska shares many characteristics with the Himalayas, including the importance of snow and glaciers in the hydrology, a large spatial extent, and a relatively low density of station observations. CCHF has also been applied to catchments in the Karakoram and Himalaya ranges, such as Hunza in the upper Indus basin. In these applications, CCHF has been used to understand snow and glacier contributions to the availability of water, the impacts of these contributions on hydropower potential, and how these contributions are projected to change with climate change. In this work, CCHF was expanded to represent glacier dynamics and the linkages between deposition of black carbon and heat transfer and melt of snow and ice.

References

Mosier, T., D. F. Hill, and K. V. Sharp. 2016. "How Much Cryosphere Model Complexity Is Just Right? Exploration Using the Conceptual Cryosphere Hydrology Framework." *The Cryosphere* 10 (5): 2147. doi:10.5194/tc-10-2147-2016.

Qian, Y., T. J. Yasunari, S. J. Doherty, M. G. Flanner, W. K. M. Lau, J. Ming, H. Wang, M. Wang, S. G. Warren, and R. Zhang. 2015. "Light-Absorbing Particles in Snow and Ice: Measurement and Modeling of Climatic and Hydrological Impact." *Advances in Atmospheric Sciences* 32 (1): 64–91. doi:10.1007/s00376-014-0010-0.

Zhan, Z.-H., J. Zhang, Y. Li, and H. S.-H. Chung. 2009. "Adaptive Particle Swarm Optimization." *IEEE Transactions on Systems, Man, and Cybernetics, Part B (Cybernetics)* 39 (6): 1362–81. doi:10.1109/TSMCB.2009.2015956.

APPENDIX E

Flood Risk Results

Map E.1 (maps E.1 and E.2 are accessible in the Nontechnical Summary, appendix A, at https://openknowledge.worldbank.org/handle/10986/35600) shows the estimated number of flood-affected people (in millions) in South Asia as of 2010. Map E.2 shows the flood-affected population overlaid with our simulation results for the Himalaya, Karakoram, and Hindu Kush (HKHK) region (chapter 5). The following indicators are exhibited for the Indus, Ganges, and Brahmaputra basins: (a) rain that runs off immediately, (b) total water production or runoff, and (c) water released from snowpack. Similar to the indicators of glacier melting, the flood-affected population in these three basins, notably the Ganges and the Brahmaputra, amounts to about 78 million people. The Indus basin is also a high-risk area, with about 15 million people estimated to be affected by floods. Thus, addressing climate change and managing black carbon are crucial to mitigating the region's potential flood risk.

Reference

Winsemius, H. C., B. Jongman, T. I. E. Veldkamp, S. Hallegatte, M. Bangalore, and P. J. Ward. 2015. "Disaster Risk, Climate Change, and Poverty: Assessing the Global Exposure of Poor People to Floods and Droughts." Policy Research Working Paper WPS 7480, World Bank, Washington, DC.

ECO-AUDIT

Environmental Benefits Statement

The World Bank Group is committed to reducing its environmental footprint. In support of this commitment, we leverage electronic publishing options and print-on-demand technology, which is located in regional hubs worldwide. Together, these initiatives enable print runs to be lowered and shipping distances decreased, resulting in reduced paper consumption, chemical use, greenhouse gas emissions, and waste.

We follow the recommended standards for paper use set by the Green Press Initiative. The majority of our books are printed on Forest Stewardship Council (FSC)–certified paper, with nearly all containing 50–100 percent recycled content. The recycled fiber in our book paper is either unbleached or bleached using totally chlorine-free (TCF), processed chlorine–free (PCF), or enhanced elemental chlorine–free (EECF) processes.

More information about the Bank's environmental philosophy can be found at http://www.worldbank.org/corporateresponsibility.

green press INITIATIVE